PROJECTS FOR THE Young Mechanic

OVER 250 CLASSIC INSTRUCTIONS & PLANS

POPULAR MECHANICS CO.

Bibliographical Note

Projects for the Young Mechanic: Over 250 Classic Instructions & Plans, first published by Dover Publications, Inc., in 2013, is a republication of a selection of material from the following works published by Popular Mechanics Co., Chicago: *The Boy Mechanic, Book 2* (1915), *The Boy Mechanic, Book 3* (1919), and *The Boy Mechanic, Book 4* (1924).

International Standard Book Number

ISBN-13: 978-0-486-49117-2
ISBN-10: 0-486-49117-X

Manufactured in the United States by Courier Corporation
49117X01   2013
www.doverpublications.com

CONTENTS

Four-Passenger Coasting Bobsled . . . 1

An Ice Boat and Catamaran 4

Mind-Reading Effect with Cards 6

An Air Pencil to Make Embossed
Letters 6

An Endless Dish or Floor Mop 6

Combination Tie Rack and Collar
Holder 7

Skates Made of Wood 7

An Ice Glider. 8

Prony Brake for Testing Small
Motors 9

A Mystic Fortune Teller 9

Holding Prints in a Liquid-Filled
Tray 10

Cellar-Door Holder 11

An Emergency Pencil Compass 11

Renewing Carbon Paper 11

How to Clinch a Finishing Nail 11

To Prevent Washbasin Bottom from
Wearing Out 11

Snowshoes: How to Make and Use
Them (Part I) 12

Soldering and Riveting 17

A Whistle 18

Card-and-Coin Trick 18

How to Make a Costumer 19

Window Catch Used for Locking an
Extension Table 19

Relieving Pressure on Heated
Canned Foods for Opening 19

Clothespin Bag 19

Snowshoes: How to Make and Use
Them (Part II). 20

Combination Settee Rocker and
Cradle. 23

A Snowball Thrower 24

Springs on the Chains of a Porch
Swing 24

Homemade Water Meter 25

A Snowball Maker 25

An Inexpensive Bobsled 26

Motor Made of Candles 26

Kettle-Handle Support 26

How to Make a Monorail Sled 27

Binding Magazines 27

A Shellac Cement 27

A Blackboard for Children 28

How to Make a Ski Staff. 28

A Game Played on the Ice. 29

An Electric Display for a Show
Window 29

Strainer for a Milk Pail 30

Baking Bread in Hot Sand 30

How to Make Small Cams 30

Display Holder for Coins 30

Holder for Skates while Sharpening . 31

A Homemade Direct-View Finder
for Cameras 31

A Non-Rolling Spool 32

How to Make a Cartridge Belt 32

Removing Iodine Stains 32

Bed-Cover Fasteners. 32

Collar Fasteners 33

Operating a Bathroom Light
Automatically. 33

A Finger-Ring Trick 33

Preventing Marks from Basting
Threads on Wool 33

House Made of Poles 35

An Electric Water Heater 39

Camps 40

A Drinking Tube 42

Washing Photographic Prints 42

Camp Furnishings 43

A Fruit Stemmer 44

A Homemade Fountain Pen 44

Destroying Caterpillars on
Grapevines 44

A Camp Provision Box 45

Wall Pockets in a Tent 47

Camp Stoves 47

Attractor for Game Fish 47

Simple Photographic-Print Washer . . 47

How to Make an Electric Fishing
Signal 48

A Chair Swing 48

Another Broom Holder 49

Squaring Wood Stock 49

A Wind Vane 49

How to Make a Flutter Ring 50

A Kitchen Utensil Hanger 50

Homemade Hinges for Boxes 50

To Remove Odors from Ice Boxes . . 50

Preventing Window Sash from
Freezing to the Sill 51

A Hanger for the Camp 51

Locking Several Drawers with
One Lock 51

A Lightning-Calculation Trick 51

An Adjustable Nutcracker 51

Substitute for a Rubber Stamp 52

A Furniture Polish 52

A Hanging Vase 52

Filing Soft Metals 52

Locking Screws in Door Hinges 52

To Remove Grease from Clothing . . 52

Paper Glider That Loops the Loop . . 53

A Water Filter 53

A Combination Electrically Operated
Door Lock 54

Lock for a Fancy Hairpin 54

An Aeroplane Kite 55

Distilling Apparatus for Water 56

Telephone Stand for a Sloping Desk . 56

The Tricks of Camping Out: Part I—
The Camping Outfit 57

The Tricks of Camping Out: Part II—
Cooking in the Woods 65

A Camp Water Bag 70

A Mold for Making Hollow Candy
Figures 70

Ornamenting an Old Tree Stump . . . 71

Marker for a Hat 71

A Quickly Made Door Latch 71

An Electric Lure for Fish 71

A Table Box for Campers 72

Novel Homemade Picture Frames . . 72

Rectangular Opening to Use over
Camera View Finder 73

Clipping File Made of Envelopes . . . 73

Handle for a Drinking Glass 73

Combination Camp-Kitchen Cabinet
and Table 74

Locking Device for Latch Hook on
Gate or Door 75

A Vanishing-Cuff Parlor Trick 75

Inexpensive Table Lamp Made of
Electrical-Fixture Parts 75

Wire Holders Keep Cabinet Doors
Open 75

"Switchboard" Protects Milker from
Cow's Tail 76

Reflected-Light Illumination with
Homemade Arrangment 76

Bedroom Shade and Curtains
Arranged for Thorough
Ventilation 76

Coffee Grinder Repaired with Rubber
Faucet Plug 77

Coal Hod Made from Iron Pipe 77

Frayed Shoe Laces Repaired with
Pitch or Wax 77

An Enameled Armchair Made of
  Wooden Strips 77

A Curling-Iron Heater. 78

A Stepmother for Incubator Chicks . 78

A Cardboard Writing and Drawing
  Pad 78

Homemade Shoulder-Pack Tent 79

Kitchen for Hikers. 80

Bird House Made of Kegs. 82

A Drinking-Glass Holder 82

Needle Threader for a Sewing
  Machine 82

Winding Coiled Springs 82

Revolving-Wheel Ruling Pen 82

A Portable Folding Boat 83

Nontangling Pasture Stake 84

Inkstand Made of a Sheet of Paper . . 84

How to Wind Wire on Electrical
  Apparatus 84

Hourglass Sewing Basket 85

A Perpetual-Motion Puzzle. 86

How to Transfer Drawings 86

Pivoted Searchlight Made of an
  Old Milk Strainer. 87

Gravity-Feed Coal Hopper on
  Truck 88

Taking Photographs in Falling
  Snow 88

A Double-Contact Vibrator 88

Battery Buzzer Converted into a
  Telegraph Sounder 89

Lawn Seats Built on Tree Stumps. . . 89

A Hinged Box Cover Made without
  Hinges 89

Automatic Flash Light Snaps
  Chicken-Coop Marauder 90

A Fishing-Tackle Outfit in a Shotgun
  Shell 90

A Split-Bamboo Lettering Pen 90

How to Make a Houseboat 91

Wood Box with a Refuse-Catching
  Drawer 92

Disappearing-Coin Trick. 92

Watering Window-Box Flowers 92

How to Make Combined Kites:
  Part I—A Dragon Kite 93

A Movable Sunshade and Seat for
  Garden Workers 96

To Keep Grass and Weeds Out of
  Tennis Courts 97

Buttonhole Bouquet Holder to Keep
  Cut Flowers Fresh 97

Cooler for a Developing Tray 97

Doorbell Push Button on Screen
  Door 98

Working Wood by the Application of
  Heat 98

A Parlor Table 99

Homemade Fuses for Battery
  Circuits. 99

Reproducing Flowers and Leaves in
  Colors. 100

Dishwasher and Drier 100

How to Make Pop-Corn Cakes 101

Threading a Darning Needle. 101

A Fish Scaler. 102

A Disappearing Towel. 102

Ammonia-Carrying Case for Insect
  Bites 102

How to Make Combined Kites:
  Part II—A Festooned Kite 103

Simple Experiment in
  Electromagnetism 104

Double Lock for a Shed 105

Ferrules for Tool Handles 105

Mallet Made from Wagon-Wheel
  Felly and Spoke 105

A Mystery Sounding Glass 105

Repairing a Broken Canoe Paddle . 106

Tightening Lever for Tennis Nets . . 106

A Desk Watch Holder 106

Cleaning Silverware 106

An Eight-Pointed Star Kite 107

Second Handle on Hoe or Rake
Saves Stooping. 108

Photo-Copying Lens Increases Angle
of Camera. 108

Belt for Sprocket Drive Made of
Brass Strips 108

Rain Alarm with Drop-of-Water
Contact 109

Coaster Steering Gear Made from
Cream-Freezer Drive 109

Pad for Glass Vessels Made of
Corks 109

A Shaving Lamp and Mirror for the
Camp 110

Automatic Electric Light on Talking-
Machine Cabinet 110

Device for Suspending Parcels from
Overhead Hooks. 110

Steel Wool as Aluminum-Ware
Cleaner 110

A Submarine Camera 111

The Magic of Numbers 114

Muffling the Ticking of a Watch or
Clock 115

A Simple Cipher Code 116

Cheese Grater and Ash Tray Made
from a Tin Can 117

An Improvised Typewriter Desk . . . 117

An Inexpensive Imitation Fire 118

Jardinière Made of Metal-Lamp
Body. 118

Replacing a Broken Coffeepot
Knob. 118

Homemade Magnesium Printer . . . 118

Decorative Toys and Boxes Made at
Home 119

Care and Storage of Camp
Equipment 124

Useful Periscope Which a Boy
Can Make. 125

Trap for Coyotes 126

Utilizing an Empty Paste Pot 126

A Colonial Mirror Frame 126

A Turntable Stand for Potted
Flowers. 128

Index for Popular Mechanics
Magazine on Bookshelf 128

Trimming Board with Foot Control
and Counterweight. 128

Folding Card Table Handy for
Invalid in Bed. 128

A Pigeon House. 129

Roses Tinged Blue by Chemicals . . 132

Making Photographic Trays 132

Camp Lantern Made of a Tin Can. . 132

How to Build Model Airplanes 134

Simple Hanger for Cut Films 136

Registers Used as Ventilators 136

Testing for Leaky Tire Valves 136

Neat Tips on Wires 136

A Neat Radio Interrupter. 137

Keeping Water Out of Automobile
Casings. 137

An Improvised Dark Room. 137

Feed Box That will not Overrun . . . 137

Homemade Greeting Cards. 138

Ventilating Doors for Garage 139

Renewing a Radio Crystal 139

Improving the Hand Drill Press . . . 139

A Homemade Motion-Picture
Camera. 140

A Practical Sliding Gate 145

An Improvised Developing Tray . . 145

Tool for Removing Headlamp
Rims. 145

A Simple Battery Radiophone 146

Useful Hints for the Motorist 147

Simple Method of Making
Candles. 148

A Simple Picture-Frame Clamp . . . 148

"Moving-Picture" Toy for Children 148

A Substantial Radiant Electric Heater 149

Book Ends Made of Angle Iron . . . 149

Dead-End Switch for Inductances . 149

Waterproofing Felt Hats 149

Motor Wheel Serves as Home-Shop Power Plant 150

Starting a Saw Cut Smoothly 152

Joining Rag Strips for Weaving . . . 152

Auxiliary Piping Prevents Overheating 152

Repair for Broken Eyeglass Spring . 152

Match-Box Holder and Lighter 153

A Two-Handled Cord Reel 153

Pulling Light Car Out of Mud 153

Keeping Insects from Light Bowls . 153

Making Varicolored Papers for Stage Effects 154

Dining Table Made from a Small Gate-Leg Table 154

How to Bleach Beeswax 154

Shooting Gallery for Toy Pistols . . . 155

How to Bake a Fish in Clay 155

Swing Stand Used during the Winter 155

A Simple Wire Stretcher 155

Making Loud Talkers for Radio-Receiving Sets 156

Efficiency of Spark-Plug Points Improved by Sharpening 156

How to Make Artificial Pearls 157

Cork Prevents Double Exposures . . 159

An Auxiliary Seat for the Auto 159

Socket Wrench for Connecting Rods 159

Thought Transference with Dice . . . 159

Reading Lamp Made from Music Rack 160

A Camp-Site Incinerator 160

Sewing Kit Made from a Cigar Box 160

A Novel Photo-Print Washer 161

The Care of Leather-Bound Books . 161

Fusible Alloys for Setting Crystals . 162

Making a Toy Catapult 162

Straightening a Bent Auto Axle . . . 162

Windmill Controlled from the House 162

Wheatstone Bridge for Coil Winding 163

Spreader for Split Rims 163

Two Effective Card Tricks 164

How to Make an Inlaid Checkerboard 165

Safety Spring-Hanger Bolts 166

A Simple Dimming Switch 166

Combined Spring Spreader and Tire Remover 166

Spring Prevents Breaking of Fishline 167

A Jumping-Frog Toy 167

The "X-Ray" Pack of Cards 167

Corked Bottle Used as Float 167

Artistic Lamps Made from Vases and Drain Tile 168

Soil Crusher Made from Mower Wheels 168

Hook Pulls Straw from Stack 168

Flexible Liquid Court-Plaster 168

FOUR-PASSENGER COASTING BOBSLED

By R. H. ALLEN

COASTER bobs usually have about the same form of construction, and only slight changes from the ordinary are made to satisfy the builder. The one shown has some distinctive features which make it a sled of luxury, and the builder will pride himself in the making. A list of the materials required is given on the opposite page. Any wood may be used for the sled, except for the runners, which should be made of ash.

Shape the runners all alike by cutting one out and using it as a pattern to make the others. After cutting them to the proper shape, a groove is formed on the under edge to admit the curve of a $\frac{5}{8}$-in. round iron rod about $\frac{1}{4}$ in. deep. The iron rods are then shaped to fit over the runner in the groove and extend up the back part of the runner and over the top at the front end. The extensions should be flattened so that two holes can be drilled in them for two wood screws at each end. If the builder does not have the necessary equipment for flattening these ends, a local blacksmith can do it at a nominal price. After the irons are fitted, they are fastened in place.

The top edges of the runners are notched for the crosspieces so that the top surfaces of these pieces will come flush with the upper edges of the runners. The location of these pieces is not essential, but should be near the ends of the runners, and the notches of each pair of runners should coin-

Coasting Is One of the Best Sports a Boy Enjoys during Winter, and a Sled of Luxury Is Something to Be Proud of among Others on a Hill or Toboggan Slide

cide. When the notches are cut, fit in the pieces snugly, and fasten them with long, slim wood screws. Small metal

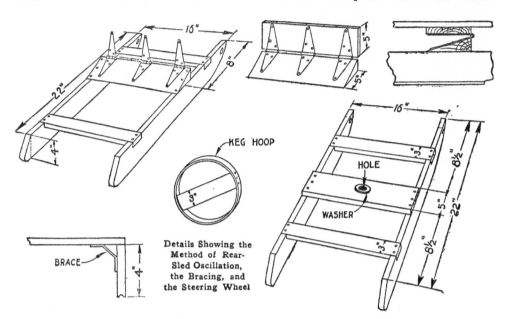

Details Showing the Method of Rear-Sled Oscillation, the Bracing, and the Steering Wheel

braces are then fastened to the runners and crosspiece on the inside, to stiffen the joint.

As the rear sled must oscillate some, means must be provided for this tilting motion while at the same time preventing sidewise turning. The construction used for this purpose is a hinged joint. The heavy 2 by 5-in. crosspiece is cut sloping on the width so that it remains 2 in. thick at one edge and tapers down to a feather edge at the opposite side. This makes a wedge-shaped piece, to which surface the three large hinges are attached. The piece is then solidly fastened to the upper edges of the runners that are to be used for the rear sled, and so located that the center of the piece will be 8 in. from the front end of the runners.

The supporting crosspiece on the front sled is fastened on top of the runners, at a place where its center will be 11 in. from the front end of the runners.

The top board is prepared by making both ends rounding and planing the surfaces smooth. On the under side, the two crosspieces are placed, which should have two ½-in. holes bored through the width of each, near the ends, to receive the eyebolts. They are placed, one with its center 12 in. from the end to be used for the rear, and the other with its center 8 in. from the front end, and securely fastened with screws. The shore is placed in the center of the board, and wires are run over it connecting the eyebolts. The eyebolts are then drawn up tightly to make the wire taut over the shore. This will prevent the long board from sagging.

On the upper side of the board and

LIST OF MATERIALS

1 top, 6½ ft. long, 16 in. wide, and 1¼ in. thick.
4 runners, 22 in. long, 4 in. wide, and 1 in. thick.
4 crosspieces, 16 in. long, 3 in. wide, and 1 in. thick.
3 pieces, 16 in. long, 5 in. wide, and 2 in. thick.
1 piece, 16 in. long, 5 in. wide, and 1 in. thick.
1 shore, 16 in. long, 3 in. wide, and 1 in. thick.

4 seat backs, 12 in. long, 16 in. wide, and 1 in. thick.
1 dowel, 3 ft. long, and 1 in. in diameter.
4 rods, ⅝ in. in diameter, and 30 in. long.
4 eyebolts, ½ in. by 6 in. long.
3 hinges, 5-in. strap.
8 hinges, 3-in. strap.

beginning at the rear end, the backs are fastened at intervals of 18 in. They are first prepared by rounding the cor-

keg hoop. A piece of wood is fastened across its diameter, and the hoop is covered with a piece of garden hose

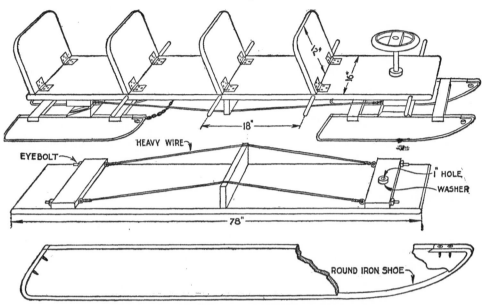

The Top Board is Well Braced on the Under Side and Fitted with Four Backs on Top to Make It a Luxurious Riding Sled, and the Runners are Provided with Metal Shoes for Speed

ners on the ends used for the tops, and the opposite ends are cut slightly on an angle to give the back a slant. They are then fastened with the small hinges to the top board. On the edges of the top board, 1-in. holes are bored about 1 in. deep, and pins driven for foot rests. These are located 18 in. apart, beginning about 5 in. from the front end. The dowel is used for the pins, which are made 4 in. long.

The steering device consists of a broom handle, cut to 18 in. in length, with one end fastened in a hole bored centrally in the 5-in. crosspiece of the front sled. A hole is bored in the top board through the center of the crosspiece fastened to the under side for the steering post. The broomstick is run through this hole after first placing two metal washers on it. After running the stick through, a hardwood collar is fastened to it just above the top board, so that the top cannot be raised away from the sled. At the upper end of the broomstick a steering wheel is attached, made from a nail-

and wrapped with twine. In the center of the crosspiece, a hole is bored to snugly fit on the broom handle, which is then fastened with screws.

The rear sled is fastened to the top board with screws through the extending wings of the hinges and into the crosspiece. Holes are bored in the front ends of all runners, and a chain or rope is attached in them, the loop end of the rear one being attached to the under side of the top board, and the one in the front used for drawing the sled.

To Prevent Drill from Catching As It Passes through Metal

The regular slope of a drill will cause the cutting edge to catch as it breaks through the metal on the opposite side of the piece being drilled. But if a twist drill is ground more flat like a flat drill, it will not "grab" into the metal as it passes through.—Contributed by James H. Beebee, Rochester, N. Y.

An Ice Boat and Catamaran

By ROBERT K. PATTERSON

THIS combination is produced by using the regular type of ice boat and substituting boats for the runners, to make the catamaran.

In constructing the ice boat, use two poles, or timbers, one 16 ft. and the other 10½ ft. long, crossed at a point 2½ ft. from one end of the longer timber. The crossed pieces are firmly braced with wires, as shown.

The mast, which should be about 12 ft. long, is set into a mortise cut in the long timber, 15 in. from the front end, and is further stabilized by wires, as shown. A jib boom, about 6 ft. long, as well as a main boom, which is 11½ ft. long, are hung on the mast in the usual manner.

The front runners consist of band-iron strips, 18 in. long, 3 in. wide, and ⅛ in. thick, with one edge ground like the edge of a skate, and the ends rounding, which are fastened with bolts to the sides of wood pieces,

The Ice Boat Provides an Ideal Outing in Winter Where There Is a Body of Water Large Enough for Sailing

18 in. long, 6 in. wide, and 2 in. thick, allowing the ground edge to project about 1 inch.

When the ice-boat frame is made of poles, the runners are attached to a piece of wood, 12 in. long, shaped as shown and fastened at right angles with bolts running through the shouldered part diagonally. This makes a surface on which the pole end rests and where it is securely fastened with bolts. If squared timbers are used, the runners can be fastened directly to them. The rear, or guiding, runner is fastened between two pieces of wood, so that its edge projects; then it is clamped in a bicycle fork, which should be cut down so that about 3 in. of the forks remain. A hole is bored through the rear end of the long pole to receive the fork head, the upper end of which is supplied with a lever. The lever is attached to the fork head by

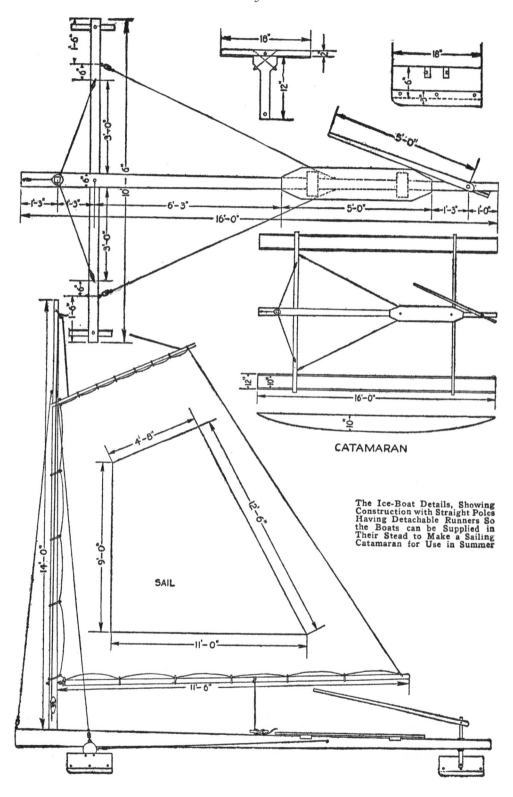

CATAMARAN

SAIL

The Ice-Boat Details, Showing
Construction with Straight Poles
Having Detachable Runners So
the Boats can be Supplied in
Their Stead to Make a Sailing
Catamaran for Use in Summer

boring a hole through the lever end at a slight angle to fit the head, allowing sufficient end to be slotted, whereupon a hole is bored through the width of the handle, and a bolt inserted, to act as a clamp.

A board is fastened on two crosspieces mortised in the upper part of the pole, for a place to sit on when driving the boat. The sail can be constructed of any good material to the dimensions given.

To rig up the ice boat for use as a catamaran, place a pole across the stern, the length of the pole being equal to the one used on the front part of the ice boat. Two water-tight boats are constructed, 16 ft. long, 12 in. wide, and 10 in. deep at the center. To make these two boats procure six boards, 16 ft. long, 10 in. wide, and 1 in. thick. Three boards are used to make each boat. Bend one board so that it will be in an arc of a circle, then nail on the two side boards, after which the edges of the sides are cut away to the shape of the bent board. The runners are removed from the ice boat, and the boats fastened to the pole ends. A rudder is attached in the place of the rear, or guiding, runner. The tops of the boats, or floats, can be covered and made water-tight.

Mind-Reading Effect with Cards

Five cards are shown, and some one person is asked to think of two cards in the lot, after which the performer places the cards behind his back and removes any two cards, then shows the remaining three and asks if the two cards in mind have been removed. The answer is always yes, as it cannot be otherwise.

To prepare the cards, take any 10 cards from the pack and paste the back of one card to another, making five double cards. Removing any two cards behind the performer's back reduces the number of cards to three, and when these are turned over they will not have the same faces so that the ones first seen cannot be shown the second time even though all five cards were turned over and shown.

An Air Pencil to Make Embossed Letters

The device illustrated is for making embossed letters on show cards, signs, post cards, etc. A small bulb, such as

The Oilcan Spout Is the Reservoir to Hold the Paint, and the Bulb Produces the Air Pressure

used on cameras, is procured, also the spout from a small oilcan. The bulb is fastened to the spout as shown.

The material for use in the pencil is quick-drying mucilage thickened with flake white. If some special color is desired, tint the mixture with aniline. Fill the spout with the mixture and attach the bulb. Squeeze the bulb gently while forming the letters, then dust over with bronze, and allow to dry.

An Endless Dish or Floor Mop

A good way to use up cord that collects about the house, is to make an endless dish or floor mop of it. Procure a thin board that will make a good length and wind the cord around it, then remove it from the board and tie the bunch together in the center.

Combination Tie Rack and Collar Holder

An unusual though simple tie rack can be made by supporting the tie bar in the center. By this arrangement the ties can be placed on it from either end,

Collar and Tie Rack with Open-End Hangers So That the Articles can be Slipped On Easily without being Passed behind a Bar as Is Usually the Case

thus avoiding the tedious threading through, required on the ordinary rack supported at each end. Collars may be hung on a peg placed above the tie bar.

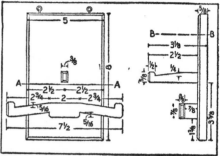

The pieces can be glued together and a good finish given in the usual way. The rack can be hung up by two screw eyes. The material required consists of four pieces, dimensioned $\frac{5}{8}$ by 5 by 8 in., $\frac{3}{8}$ by $\frac{7}{8}$ by $7\frac{1}{2}$ in., $\frac{3}{8}$ by $\frac{5}{8}$ by $3\frac{1}{8}$ in., and $\frac{7}{8}$ by $\frac{7}{8}$ by 2 in. respectively.—Contributed by Arthur C. Vener, Dallas, Texas.

Skates Made of Wood

Skates that will take the place of the usual steel-runner kind and which will prevent spraining of the ankles, can be made of a few pieces of $\frac{1}{2}$-in. hardwood boards.

Four runners are cut out, 2 in. wide at the back and $1\frac{1}{2}$ in. wide at the front, the length to be 2 in. longer than the shoe. The top edges of a pair of runners are then nailed to the under side of a board 4 in. wide, at its edges.

A piece of board, or block, 2 in. wide is fastened between the runners at the rear, and one 1 in. wide, in front. Two bolts are run through holes bored in the runners, one just back of the front board, or block, and the other in front of the rear one.

Four triangular pieces are fastened, one on each corner, so that the heel and toe of the shoe will fit between them, and, if desired, a crosspiece can

be nailed in front of the heel. Straps are attached to the sides for attaching

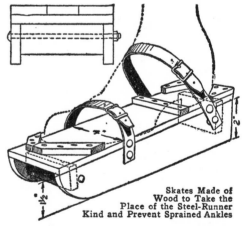

Skates Made of Wood to Take the Place of the Steel-Runner Kind and Prevent Sprained Ankles

the skate to the shoe. Both skates are made alike.—Contributed by F. E. Kennar, Hennessey, Okla.

❪The best paint for paper roofing is asphaltum varnish.

An Ice Glider

By MILDRED E. THOMAS

THE enthusiastic pushmobilist need not put aside his hobby during the winter, as an amusement device for use on ice, which will surpass the very best pushmobile, can be easily made as shown in the illustration.

Similar to an ice yacht, only a great deal smaller, the ice glider will require three ordinary skates, two of which are fastened to the ends of the front crosspiece, so that their blades will stand at an angle of about 30 deg. with their edges outward. To get this angle, tapering blocks are fastened to the crosspiece ends, as shown. The skates are then fastened to these blocks.

The crosspiece is 30 in. long and about 8 in. wide. In the center of this piece an upright is constructed, 26 in. high. The edges of the front crosspiece are cut on a slant so that a piece nailed to its front and back edge will stand sloping toward the rear. A handle, 24 in. long, is fastened between the two uprights at the upper end. The rear part is made of a board, 8 in. wide and 40 in. long. The remaining skate is fastened in a perfectly straight position on the rear end. The skates may be attached with screws run through holes drilled in the top plates, or with straps. The front end of the rear board has a hole for a bolt to attach it to the center of the front crosspiece, so that the latter will turn to guide the glider.

A pusher is prepared from a block of wood, into which nails are driven with their ends projecting on the under side. The block is strapped to one shoe, as shown.

The glider is used in the same manner as a pushmobile.

The pusher can be made in another way by using sole leather instead of the block. Small slots are cut in the sides for the straps. Nails are driven

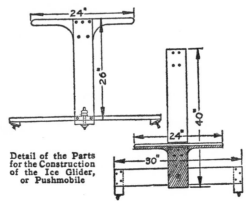

The Glider is Pushed over the Ice Similarly to a Pushmobile, and the Speed That can be Attained is Much Greater

Detail of the Parts for the Construction of the Ice Glider, or Pushmobile

24"

26"

24"

40"

30"

through the leather so that the points project. Either kind of pusher is

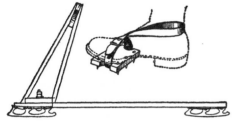

The Block of Wood with Protecting Nails to Fasten on the Shoe That Does the Pushing

especially adapted for the pushmobile to prevent wear on the shoe.

Prony Brake for Testing Small Motors

The ordinary prony brake is not, as a rule, sensitive enough to make an accurate test on small motors, such as those used in driving sewing machines, washing machines, vacuum cleaners, etc. The arrangement shown in the accompanying sketch has been used for this purpose with good results and was very accurate. The operation of the brake is exceedingly simple.

A pulley without a crown face is attached to the shaft of the motor, which

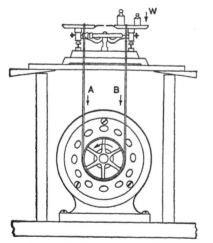

Prony Brake Used in Connection with a Small Balance to Find the Horsepower

is fastened to the top of a table or bench, and a balance mounted directly over the pulley. The support for the balance should be a narrow strip, which

in turn is supported on two upright pieces, as shown. A light rope is put under the pulley, and the ends are looped over the platforms of the balance so that it does not interfere with the operation of the balance. The ends of the rope should be vertical and parallel. The piece upon which the balance rests is raised by inserting wedges, thus increasing the tension in the rope. The resulting friction of the rope on the pulley increases the load.

If the motor is running in the direction indicated by the arrow on the pulley, the tension in the left-hand end of the rope will be greater than in the right-hand end and a weight must be placed on the right-hand platform of the balance. When the weight W is adjusted so that the two pointers on the platforms are exactly opposite each other, the value of the weight W, in pounds, will represent the difference in pull, in pounds, between A and B. If the value of the weight W is known and also the speed of the machine when the weight was determined, the horsepower output can be computed by means of the following equation:

$$Hp. = \frac{6.2832 \times L \times W \times R.P.M.}{33,000 \times 12}$$

In this equation, L is the distance in inches from the center of the pulley to the center of the rope. Two ordinary spring balances may be substituted for the beam balance and the difference in their readings taken for the value W. For best results, the tension in the slack end of the rope should be as small as possible, and it may be necessary to wrap the rope one or more times completely around the pulley.

A Mystic Fortune Teller

Fortune telling by means of weights striking glasses or bottles is quite mysterious if controlled in a manner that cannot be seen by the audience. The performer can arrange two strikes for "no," and three for "yes" to answer questions. Any kind of bottles, glass, or cups may be used. In the bottles the pendulum can be suspended from the cork, and in the glasses from small tripods set on the table.

The secret of the trick is as follows: A rubber tube with a bulb attached to

Holding Prints in a Liquid-Filled Tray

After having considerable trouble in keeping my paper prints in the hypo fixing bath from curling, which would

BULB

BULB

The Rocking of the Table is Caused by the Pressure of Air in the Bulb under the Foot, the Movement Causing the Pendulum to Swing and Strike the Glass

each end is placed under a rug, one bulb being located under one table leg and the other near the chair of the performer set at some distance from the table where it can be pressed with the foot. Some one selects a pendulum; the performer gazes intently at it, and presses the bulb under his foot lightly at first; then, by watching the swaying of the pendulum selected, he will know when to give the second impulse, and continue until the weight strikes the glass. As the pendulums are of different lengths they must necessarily swing at different rates per second. The impulses must be given at the proper time or else the pendulum will be retarded instead of increased in amplitude. A table with four legs is best to use, and the leg diagonally opposite that with the bulb beneath it must not touch the carpet or floor. This can be arranged by placing pieces of cardboard under the other two legs. —Contributed by James J. McIntyre.

force the edges out of the liquid, I found the plan here illustrated a success. I procured a piece of wood, the size of a postcard, and stuck four glass push pins into one surface, one at each corner, and fastened a handle to the center of the upper side. The papers are first placed in the bath, then

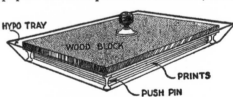

HYPO TRAY

WOOD BLOCK

PRINTS

PUSH PIN

Push Pins on the Under Side of the Board Raise It and Provide a Space for the Prints in the Liquid

the board is set over them with the pins down. This holds the prints under the liquid but does not press them tightly together.—Contributed by J. J. Kolar, Maywood, Ill.

❡A piece of an old gunny sack will polish brass work very nicely.

Cellar-Door Holder

A cellar door that opened up against a wall required a catch of some kind to keep it open at times. As I did not

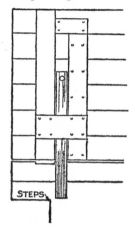

want a catch to show on the wall, I devised a holder as shown. Three pieces of wood were nailed to the under side of the door in such positions that they formed a recess in which a fourth piece, 2 in. wide and 1 in. thick, would slide endways. A knob was attached to the upper end of the slide, which served the double purpose of a handle and a stop for the slide. The manner of using the holder is clearly shown.—Contributed by H. T. Smith, Topeka, Kansas.

An Emergency Pencil Compass

The need of a compass when none was at hand caused me to quickly devise a substitute for the work. A piece of stiff wire, about the length of the pencil, was procured, and several turns were made around the pencil, as shown. The lower straight end was filed to a point. The wire can be bent to obtain the radius distance.—Contributed by Preston Ware, Rome, Ga.

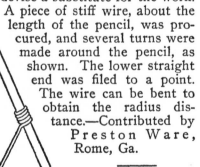

⁅A very effective dip for brass and copper articles, that will leave a clean and bright finish, is 2 qt. of aqua fortis, 1 gal. of sulphuric acid, 1 pt. of water and a pinch of salt.

Renewing Carbon Paper

When carbon paper has been used several times, the preparation becomes almost worn off on some parts, while other parts of the paper are as good as new. The process of renewing is very simple and it can be done by anyone without special apparatus. All that is necessary is to hold the paper in front of a fire or over a radiator a few seconds. The heat will cause the preparation to dissolve and spread over the paper, so that when it is dry the paper will have a new coating. This can be repeated, and in some cases will double the life of the carbon paper.—Contributed by Chester M. Kearney, Danville, Quebec.

How to Clinch a Finishing Nail

A wire or finishing nail may be clinched as nicely as a wrought nail,

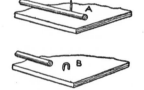

if a nail punch or piece of iron is placed along the side of it, as shown at A, and the nail hammered into an arched form, as at B. The punch or rod is then withdrawn and the arch driven into the wood.—Contributed by James M. Kane, Doylestown, Pa.

To Prevent Washbasin Bottom from Wearing Out

The ears from some sirup buckets were removed and three of them soldered, at equal distances apart, on the bottom of the washbasin near the outside edge of the lower part. These prevented the wear from coming on the bottom of the basin, and it lasted several times as long as ordinarily.—Contributed by A. A. Ashley, Blanket, Texas.

⁅To curl feathers, heat slightly before a fire, then stroke with something like the back of a case knife.

SNOWSHOES

How to Make and Use Them

By Stillman Taylor

PART I—Shapes of Snowshoes

T O the inventive mind of the North American Indian we owe the snowshoe, and its conception was doubtless brought about through that prolific source of invention—necessity. The first models were crude web-footed affairs, but improvements in model and manner of filling the frames were gradually added until the perfected and graceful shoe of the present was finally reached. The first snowshoes were made by the Indians, and the Indians of Maine and Canada continue to fashion the finest models today.

The snowshoe is a necessity for the sportsman and trapper whose pleasure or business leads him out in the open during the winter season, when roads and trails are heavily blanketed by a deep fall of powdery snow. But the use of the web shoe is by no means confined to the dweller in the wilderness, since the charm of wintry wood and plain beckons many lovers of the outdoors to participate in this invigorating sport, and snowshoe tramps are fast growing in popularity in and about our cities and towns.

All the modern snowshoes are constructed upon practically the same general lines, although the types of frames differ considerably in size as well as in shape, and the filling of hide is often woven in many varied and intricate patterns. The frame or bow—usually made of ash in order to get strength with light weight—is bent in many shapes, but the one shown in the diagram is a typical general-purpose shoe, and may be called standard. The frame is held in shape by means of two wooden cross braces, neatly mortised into the frame. These braces are spaced some 15 or 16 in. apart, and so divide the shoe into three sections, known as the toe, center, and heel. The filling is woven into a lanyard, which is a light strip of hide firmly laced to the frame through a double row of holes drilled in the wood. The center filling is woven of heavy strands of rawhide, in a fairly coarse mesh, because this part of the shoe must bear the weight of the body and the brunt of wear. The end fillers for toe and heel are woven of lighter strands of hide, and the mesh is, of course, smaller.

As may be noted by referring to the drawing, a center opening or "toe hole" is provided, and as the greater strain on the filling lies directly under the ball of the foot, the shoe is rein-. forced at this point by the "toe cord" running across, and the "toe-cord stays," which are tied in on each side of the toe hole—one end being fastened to the toe cord and the other lashed over the wooden cross bar of the frame. These reinforcing cords are formed of several strands of hide, the stays being again wound with finer strands.

To prevent slipping and to secure a good foothold while walking, the manner of attaching the foot to the shoe is of importance, and this is done by making use of a toe strap, which will allow the toe to push down through the toe opening as the heel of the foot is lifted in the act of walking. A second strap, or thong, leading from the top around the foot, above the curve of the heel, is needed to lend additional support in lifting the snowshoe, to effect the easy shambling stride characteristic of the snowshoer.

There are, of course, a great number of models or styles, some one style being popular in one locality, while an altogether different style is preferred in another part of the country. The most representative types are well shown in the illustrations, and a brief description will point out their practical advantages, because each model possesses certain merits—one model being designed for fast traveling in the open, another better adapted for brush travel, while others are more convenient for use in a hilly country where much climbing is done, and so on.

Style A is regarded by snowshoe experts as an extreme style, for it is long and narrow. It is designed for fast traveling over smooth and level country, and over loose, powdery snow. This style is much used by the Cree Indians, and is usually made 12 in. wide by 60 in. long, with a deeply upcurved toe. It is a good shoe for cross-country work, but is somewhat difficult to manage on broken trails, when the snow is packed, and also affords rather slippery footing when crossing ice. Owing to the stout construction of the frame and reinforcement needed to retain the high, curved toe, style A is more difficult to manage than the more conservative models, and its stiffness of frame makes it more fatiguing to wear, while its use is a decided handicap in mountainous districts, because a curved toe always makes hill climbing more difficult.

Style B may be considered the ordinary eastern model, and a common style best adapted for all-around use. It is a neat and gracefully designed frame, about 12 in. wide and 42 in. long, and is usually made with a slightly upcurving toe, about 2 in. turn at the toe being correct. When made by the Indians of Maine, this model is fashioned with a rather heavy heel, which is an advantage for fast walking, while it increases the difficulty in quick turning.

Style C is a favorite model among the hunters and woodsmen of New England. This is a splendid style for general purposes in this section of the country, since the full, round toe keeps the toe up near the surface, and lets the heel cut down more than the narrow-toe models. Style C is an easy shoe to wear, and while not so fast as the long, narrow frame, its full shape is more convenient for use in the woods. It is usually made with about 1 to 1½-in. turn at the toe.

Style D is the familiar "bear's paw," a model originating with the northeastern trapper. This model is well adapted for short tramps in the brush, and having a flat toe, is likewise a good shoe for mountain climbing. For tramping about in thick brush, a short, full shoe enables one to take a shorter stride and turn more quickly, but it is a slow shoe for straight-ahead traveling.

When purchasing a pair of snowshoes, some few important considerations should be kept in mind, and the size and model will depend upon the man to some extent, since a large, heavy man will require a larger snowshoe than would suffice for a person of lighter weight. Height also enters into the choice, and while a small person can travel faster and with less fatigue when equipped with a proportionately small shoe, a tall man will naturally pick out a larger-sized snowshoe for his use. For a country where deep snows prevail, larger sizes are best, but in localities where the snow packs solidly and there is considerable ice, and in mountainous districts, or for rough-country traveling, the smaller sizes will give more satisfaction and prove more durable also. For a wet-snow locality, the center filling should be strung in rather coarse mesh, while for soft, powdery snow, a finer mesh will be the logical choice.

There are snowshoes and snowshoes, and while there are fine models regularly stocked by a few of the better sporting - g o o d s firms, there is likewise a deal of poorly made snowshoes on the market. It is well to pay a fair price and secure a d e p e n d a b l e handmade article, for the cheaper snowshoes —often filled w i t h seine twine and t h e cheapest hide (commonly known in the trade as "gut")—will warp and twist in the frame, and the shoddy filling will soon become loosened up and "bag" after a little use. The best snowshoes that the writer is acquainted with are made by the Indians, and the filling is ordinarily made of neat's hide; cowhide for the center filling, and calfskin for the toe and heel. A first-class pair of snowshoes may be had for about $6 to $7.50, and when possible to do so, it is best to have them made to order. This p l a n is, of course, necessary in case one wishes to incorporate any little wrinkles of his own into their making, or desires a flatter toe, lighter heel, or a different mesh from the usual stock models.

Where but one pair of snowshoes is purchased, style B will probably prove the best selection, and should be ordered with the flat toe, or a turn not greater than 1 in. The frame may be in either one or two pieces, depending upon the size of the shoe and the ideas of the Indian maker, but it is well to specify white ash for the frames in the order. No Indian maker would be guilty of using screws or other metal fastenings, but many of the cheap and poorly fashioned snowshoes are fastened at the heel with screws, thus making this a decidedly weak

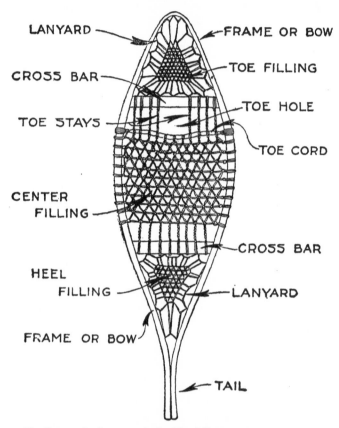

The Frame of a Snowshoe in Its Usual Construction, Showing the Crosspieces with Their Laced Fillings of Hide and the Different Parts Named, for a Ready Reference

point, since the wood is quite certain to split after a little rough service. In contrast to the poor workmanship of these low-priced snowshoes, the Indian-made article is fashioned from sound and properly seasoned wood; the cross bars are snugly fitted by mortising to the frame; the filling is tightly woven, and the heel is properly fastened by lacing with a rawhide

it is a good idea to select a filling of good heavy weight and with a firmly woven and open mesh, say, about ¾ in. The toe and heel sections will, of course, be of finer-cut hide and smaller mesh, and it is wise to avoid those shoes employing seine twine for the end filling. Some factory-made snowshoes are given a coat or two of varnish, but this, while serving to make them partly waterproof, makes them rather slippery when crossing logs and ice. Most woodsmen prefer to leave both frame and filling in their natural condition.

The Indian-made snowshoe is always

thong. However, Indian makers are likely to make the toe small and leave the wood to form a rather heavy heel. Some few woodsmen and sportsmen may prefer this model, but the majority favor a fuller toe and a lighter heel for general use, because the regulation Indian model, cutting down at toe and heel equally deep, increases the difficulty of easy traveling over soft snow, although it is a good shoe when used over broken trails.

When buying snowshoes at the store, see that the frames are stoutly and well made, and for all-around use,

provided with a generously large toe hole, so that ample foot covering may be used. This point is generally overlooked in the machine-made product, and the toe cords are also frequently r o u g h l y formed, thus chafing the feet and making them sore. These details may or may not prove a handicap for short tramps n e a r town, but for long t r i p s through the woods, they are important considerations.

The Indian manner of tying the snowshoe to the foot by means of a single twisted and knotted thong is a good method of attachment, in that, if the thong is properly ad-

justed to the requisite snugness in the first place, the shoes may be quickly removed by a simple twist of the ankle. A better fastening is secured by using a fairly wide (¾ in.) toe strap and a long thong. The toe strap is placed over the toes, immediately over the ball of the foot, and secured against slipping by weaving the ends in and out between the meshes of the filling until it reaches the frame on either side. This grips the toe strap firmly and does away with the necessity of tying a knot. A narrow thong, about 4 ft. long, is now doubled, the center placed just

above the heel of the foot, and the ends passed under the toe cord, just outside of the toe-cord stays on each side. The thong is then brought up and across the toes, one end passing over and the other under the toe strap. Each end of the thong is now looped around the crossed thong, on either side, and then carried back over the back of the heel and knotted with a common square or reef knot. Calfskin makes a good flexible foot binding, or a suitable strip of folded cloth or canvas may be used.

The regulation snowshoe harness, consisting of a leather stirrup for the toe and an instep and heel strap, will be found more comfortable than the thong, and when once adjusted snugly to the foot, the shoes may be quickly taken off and put on again by pushing the heel strap down, when the foot may be slipped out of the toe stirrup.

The use of heavy leather shoes is of course undesirable, and the only correct footwear for snowshoeing is a pair of high-cut moccasins, cut roomy enough to allow one or more pairs of heavy woolen stockings to be worn. The heavy and long German socks, extending halfway to the knee, drawn on over the trouser legs, are by far the most comfortable for cold-weather wear. The feet, thus shod, will not only be warm in the coldest weather, but the free use of the toes is not interfered with. Leather shoes are cold and stiff, and the heavy soles and heels, chafing against the snowshoes, will soon ruin the filling.

Soldering and Riveting

By JOHN D. ADAMS

There are two simple processes that every experimenter should master: soldering and riveting. The large soldering copper will find only a very re-

A Small Torch Made of a Penholder is Handy to Use in Soldering Electrical Apparatus

stricted use with the amateur on account not only of its clumsiness, but of the fact that it requires a fire, which is often impracticable to obtain. The experimenter should therefore construct a small alcohol lamp, which, after a little experience, will reveal the following advantages: It may be brought into instant use at any place; it will make a more perfect connection; with a small blowpipe places may be reached that are entirely inaccessible to the large iron; several small pieces may be set in position and soldered without disturbing them, which is quite impossible with the large iron.

To make such a lamp, procure a small wide-mouthed bottle so that very little alcohol will be necessary and the lamp may be tipped at any desired angle. A short piece of seamless brass tubing should be procured, or, preferably, one of those capped brass cylinders for holding pencil leads, the button of which should be sawn off and the cap used to keep the alcohol from evaporating. A good, sound cork is next in order, and in cutting the central hole, use the brass tube, which should be sharpened around the lower end. Proceed with a rotary motion, and a clean core will be removed. If an ordinary lamp wick is not at hand, soft cotton string may be bundled up as a substitute. Such a lamp is s a f e, odorless and will not blacken the work in the least as in the case of kerosene or gasoline.

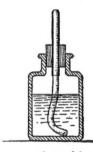

There are many good soldering fluxes on the market, but that obtained by dissolving as much scrap of zinc as possible in muriatic acid will solder practically everything that may be necessary, provided, of course, the surfaces are filed or scraped bright. Wire solder is usually the most convenient, as small pieces can be readily cut off and placed directly on the work where required. A small blowpipe is often a valuable adjunct, as it makes possible a long, narrow flame that may be directed in almost any direction.

Where numerous small connections are to be made, as is often the case with electrical apparatus, the small torch illustrated will be found very convenient. It is simply an old penholder with the wood portion shortened somewhat and the metal end filed off square and cleaned out. This is then  filled with wicking, and it is only necessary to dip it in alcohol in order to soak up enough to solder an ordinary connection.

The second simple process, of which many fail to appreciate the usefulness in experimental work, is that of riveting—particularly when done on a small scale. Very often the material in hand is tempered steel and cannot, therefore, be soldered to advantage, or it may be a case where subsequent heating makes a heat-proof connection imperative. Then, again, the joint may require the combined strength of both solder and rivet.

When properly set, the strength of the ordinary brass pin, when used as a rivet, is quite great. Should the work require a particularly soft rivet, it is only necessary to hold the pin for a moment in the flame of a match. A somewhat larger and stronger rivet may be made by softening and cutting to the required length the small flat-headed nails used in making cigar boxes. The ordinary shingle nail is also of a suitable shape after the burrs have been filed off under the head.

In setting these small rivets, it is absolutely necessary that they closely fit the holes, as at A, otherwise the result will be as indicated at B in the sketch. Be careful not to leave too great a length for rounding over on the metal. This extra length should approximately equal the diameter of the

rivet and must be filed flat on the top before riveting. In case of pins, it will be found easier to cut them off to the

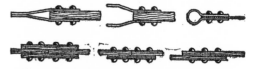

A Few Joints Where Rivets are Used to Hold the Parts Solidly Together

proper length after they are inserted. Use the smallest hammer available, striking many light blows rather than a few heavy ones.

A Whistle

Cut a circular piece of tin any convenient size, preferably 3 in. in diameter, and bend it across the diameter so 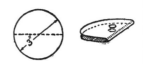 that it will be in a narrow U-shape. Then drill or punch a hole through both parts as shown. Place it in the mouth with the open edges out, being sure to press the lips on the metal tightly on both upper and lower pieces outside of the holes and to rest the tongue against the edge of the tin, even with the holes, and blow.

The result of the first attempt may not be a sound, but with a little practice any familiar tune may be whistled.—Contributed by Chas. C. Bradley, W. Toledo, O.

Card-and-Coin Trick

If a card is balanced on the finger and a coin placed on the card directly over the finger, one would not think that the card could be flipped out leaving the coin on the finger end. This is easily accomplished,  if care is taken to snap the card sharply and squarely.—Contributed by R. Neland, Minneapolis, Minn.

How to Make a Costumer

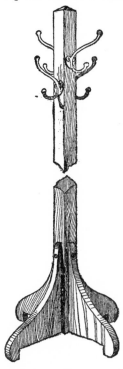

With but little skill, and such tools as are ordinarily found around a home, a plain but serviceable costumer can be made, as shown in the sketch. The necessary materials for it are: One main post, 1½ in. square and about 6½ ft. long; four legs, or foot brackets, ¾ by 6 by 9 in.; four brass clothes hooks, and the necessary screws and varnish for assembling and finishing.

The center post should be chamfered at the top to relieve the abruptness. The four legs should all be made alike and in some shape that allows them to be fastened to the post in a simple manner. In the sketch, the legs are fastened to the post by one visible screw at the top and one put in on an incline through the bottom edge of the leg. The clothes hooks are fastened to the post in pairs at different heights, thereby preventing the screws of adjacent hooks from running into one another. The finish of the costumer should be such as to match the woodwork of its surroundings.—Contributed by Harry A. Packard, Norway, Maine.

Window Catch Used for Locking an Extension Table

To prevent the two ends of an extension table from pulling apart when not desired, an ordinary window catch can be fastened and locked in place to the under side of the table top with one part on each end of the table. If but one catch is used and fastened in the center, it is best to mark it off first, and then pull the table ends apart to fasten the catch more easily. It may be desired to use two catches for a very heavy table, in which case it would be best to place one on either side of the center.—Contributed by F. M. Griswold, New York, N. Y.

Relieving Pressure on Heated Canned Foods for Opening

In opening a can of food that has been heated, the instant the cover is punctured the steam will force out a part of the contents, which is very annoying. To avoid this, pour a little cold water on the cover and allow it to remain a few seconds, then turn it off and immediately puncture the cover. This will counteract the interior force, and the can may be opened without trouble.—Contributed by Joseph Kohlbecher, Jr., San Francisco, Cal.

Clothespin Bag

Clothespins are usually kept in a bag, and the one our home possessed had a draw string which would always stick and hold the bag shut. The remedy for this, and a time saver also, was to remove the draw string and insert instead a piece of wire, which was afterward shaped to a circle with an eyelet at the joint. The bag can be hung on a nail and the mouth is always open to its fullest extent, yet lies flat against the wall.—Contributed by Jas. A. Hart, Philadelphia, Pa.

SNOWSHOES

HOW TO MAKE AND USE THEM

By Stillman Taylor

PART II—Making the Shoe

[In making the snowshoe it may be necessary to refer to the previous chapter to select the style, or to locate the name of the parts used in the description.—Editor.]

SNOWSHOE making is an art, and while few, if any, white men can equal the Indian in weaving the intricate patterns which they prefer to employ for filling the frames, it is not very difficult to fashion a good solid frame and then fill it by making use of a simple and open system of meshing. For the frames, white ash is much the best wood, but hickory and white birch are dependable substitutes, if the former cannot be obtained. Birch is perhaps the best wood to use when the sportsman wishes to cut and split up his own wood, but as suitable material for the frames may be readily purchased for a small sum, probably the majority of the readers will elect to buy the material. Any lumber dealer will be able to supply white ash, and it is a simple matter to saw out the frames from the board. The sawed-out frame is inferior to the hand-split bow, but if good, selected material can be obtained, there will be little, if any, difference for ordinary use.

When dry and well-seasoned lumber is used, the frame may be made to the proper dimensions, but when green wood is selected, the frame must be made somewhat heavier, to allow for the usual shrinkage in seasoning. For a stout snowshoe frame, the width should be about $1\frac{1}{16}$ in.; thickness at toe, $\frac{7}{16}$ in., and thickness at heel, $\frac{9}{16}$ in. The frame should be cut 2 in. longer

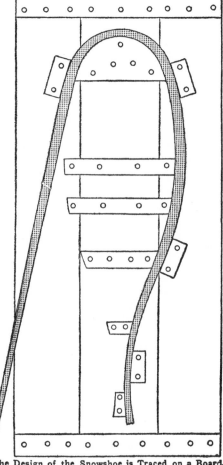

The Design of the Snowshoe is Traced on a Board, and Blocks are Used to Shape the Frame or Bow

20

than the finished length desired, and in working the wood, remember that the toe of the finished frame will be the center of the stick; the heel, the end of the stick, and the center of the shoe will lie halfway between the heel and toe.

Locate the Cross Bars by Balancing the Frame, Then Fit the Ends in Shallow Mortises

After the frames have been finished, the dry wood must be steamed before it can be safely bent to the required shape, and before doing this, a wooden bending form must be made. An easy way to make this form is to first draw a pattern of the model on a sheet of paper, cut out the pencil mark, and, placing this pattern on a board, carefully trace the design on the wooden form. A number of cleats, or blocks, of wood will now be needed; the inside blocks being nailed in position, but the outside stay blocks being simply provided with nails in the holes, so they may be quickly fastened in position when the steamed frame is ready for the form.

To make the frame soft for bending to shape, steaming must be resorted to, and perhaps the easiest way of doing this is to provide boiling water in a

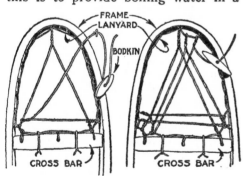

Begin Weaving the Toe Filling at the Corner of Cross Bar and Frame, Carrying It Around in a Triangle until Complete

wash boiler, place the wood over the top, and soak well by mopping with the boiling water, shifting the stick about until the fibers have become soft

and pliable. After 10 or 15 minutes of the hot-water treatment, wrap the stick with cloth and bend it back and forth to render it more and more pliable, then use the hot-water treatment, and repeat the process until the wood is sufficiently soft to bend easily without splintering. The toe being the greatest curve, must be well softened before putting on the form, otherwise the fibers are likely to splinter off at this point. When the frame is well softened, place it on the bending form while hot, slowly bend it against the wooden inside blocks, and nail on the outside blocks to hold it to the proper curve. Begin with the toe, and after fastening the outside blocks to hold this end, finish one side, then bend the other half to shape. The bent frame should be allowed to dry on the form for at least a week; if removed before the wood has become thoroughly dry and has taken a permanent set, the frame will not retain its shape. The same bending form may be used for both frames, but if one is in a hurry to finish the shoes, two forms should be made, and considerable pains must be taken to make them exactly alike in every way.

When the frames are dry, secure the tail end of the frame by boring three holes about 4 in. from the end, and fasten with rawhide. The work of fitting the two cross bars may now be undertaken, and the balance of the snowshoe depends upon fitting these bars in their proper places. Before cutting the mortise, spring the two bars in the frame about 15 in. apart, and balance the shoe in the center by holding it in the hands. When the frame exactly balances, move the bars sufficiently to make the heel about 3 oz. heavier than the toe, and mark the place where the mortises are to be cut. The cross bars and mortise must be a good tight fit, and a small, sharp chisel will enable the builder to make a neat job. It is not necessary to cut the mortise very deep; $\frac{1}{4}$ in. is ample to afford a firm and snug mortised joint.

The lanyard to which the filling is woven is next put in, by boring pairs of small holes in the toe and heel sections, and lacing a narrow rawhide thong through the obliquely drilled holes. Three holes are then bored in the cross bar—one on each side about 1½ in. from the frame, and the third in the center of the bar; the lanyard being carried through these holes in the cross bar.

Begin the toe filling first, by making an eye in one end of the thong, put the end through the lanyard loop and then through the eye, thus making a slipknot. Start to weave at the corner where the bar and frame are mortised, carry the strand up and twist it around the lanyards in the middle of the toe, then carry it down and make a like twist around the lanyard loop in the opposite corner. The thong is now looped around the next lanyard (No. 2 from the cross-bar lanyard) and fastened with the twisted loop knot illustrated. Continue the strand across the width of toe space and make a similar loop knot on No. 2 lanyard on the starting side, twist it around the strand first made and loop it under the next cross-bar lanyard loop, then carry it up and twist it around the lanyard loop in the toe of the frame, continuing in the same manner until the last lanyard of the toe is reached, when the space is finished by making the twisted loop knot until the space is entirely filled. It is a difficult matter to describe by text, but the illustrations will point out the correct way, and show the manner of making an endless thong by eye-splicing, as well as illustrating the wooden bodkin or needle used in pulling the woven strands taut. This bodkin is easily made from a small piece of wood, about ¼ in. thick, and about 2 in. long.

An Endless Thong is Made with Eyes Cut in the Ends of the Leather, and Each Part is Run through the Eye of the Other

To simplify matters, the heel may be filled in the same manner as the toe.

For the center, which must be woven strong and tight, a heavier strand of hide must be used. Begin with the toe cord first, and to make this amply strong, carry the strand across the frame five or six times, finishing with a half-hitch knot, as shown, then carry

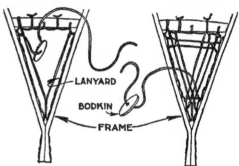

The Heel Filling is Woven by Making the Connection with the Lanyard in the Same Manner as for the Toe Filling

it up and twist it around the cross bar to form the first toe-cord stay.

As may be noted, the center section is filled by looping back and twisting the strands as when filling the toe. However, the filling is looped around the frame instead of a lanyard, and a clove hitch is used. A toe hole, 4 in. wide, must be provided for, and when enough of the filling has been woven in to make this opening, the thong is no longer looped around the cross bar, but woven through the toe cord. As the filling ends in the toe cord, it should be woven in and out at this point several times, finishing the toe hole by looping a strand around the cross bar at the side of the toe hole, then passing it down the toe-cord stay by twisting around it; then twisted around the toe cord along the filling to the other side of the toe hole, where it is twisted around the toe-cord stay on the opposite side, looped around the frame and ended in a clove hitch.

At the first reading, it will doubtless appear difficult, but a careful examination of the illustrations will soon show how the trick is done, and indeed it is really a very simple matter, being one of those things which are easier to do than it is to tell how to do them. The method of filling has been purposely made simple, but the majority of shoes are filled in practically the same manner, which answers quite as well as the more intricate Indian design.

The knack of using the snowshoe is quickly mastered, providing the shoes are properly attached, to allow the toe

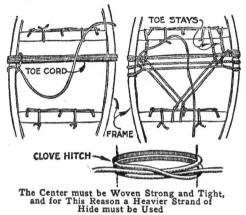

The Center must be Woven Strong and Tight, and for This Reason a Heavier Strand of Hide must be Used

ample freedom to work down through the toe hole as each foot is lifted. The shoe is, of course, not actually lifted in the air, but rather slid along the surface, half the width of one shoe covering the other when it is lifted in the act of walking. At first the novice may be inclined to think snowshoes a bit cumbersome and unwieldy, and doubt his ability to penetrate the brush. However, as the snowshoer becomes accustomed to their use, he will experience little if any difficulty in traveling where he wills. When making a trail in a more or less open country, it is a good plan to blaze it thoroughly, thus enabling one to return over the same trail, in case a fall of snow should occur in the meantime, or drifting snow fill up and obliterate the trail first made. When the trail is first broken by traveling over it once by snowshoe, the snow is packed well and forms a solid foundation, and even should a heavy fall of snow cover it, the blaze marks on tree and bush will point out the trail, which will afford faster and easier traveling than breaking a new trail each time one journeys in the same direction.

A well-made pair of snowshoes will stand a couple of seasons' hard use, or last for a year or two longer for general wear. To keep them in good shape, they should be dried out after use, although it is never advisable to place them close to a hot fire, or the hide filling will be injured. Jumping puts severe strain on the frame of the shoe, and while damage may not occur when so used in deep, soft snow, it is well to avoid the possibility of breakage. Accidents will now and then happen, to be sure, and as a thong may snap at some unexpected moment, keep a strand or two of rawhide on hand, to meet this emergency.

Combination Settee Rocker and Cradle

By fastening a frame with hinges to the front of a settee rocker, a combination piece of furniture can be made, which may be used either as a regular settee or as a cradle. For this purpose, a covered frame should be provided, being sufficiently long to extend across the front between the arm supports and having such a width that it will easily fit under the arms when hinged to the seat, as shown in the illustration. To keep the frame in position while serving as a cradle front, or when turned down for regular use, screw hooks are placed at each end, so that, in the former case, the frame, when swung up, can be secured in place by attaching the hooks to screw eyes fastened under the arm supports; while, for regular use, the frame is secured in its swung-

A Settee Rocker with a Front Attachment to Make It into a Cradle When Desired

down position by fastening the hooks into screw eyes properly placed in the front legs.—Contributed by Maurice Baudier, New Orleans, La.

A Snowball Thrower

By ALBERT BATES, Jr.

The snow fort with its infantry is not complete without the artillery. A set of mortars, or cannon, placed in the fort to hurl snowballs at the entrenched enemy makes the battle more real. A device to substitute the cannon or a mortar can be easily constructed by any boy, and a few of them set in a snow fort will add greatly to the interest of the conflict.

The substitute, which is called a snowball thrower, consists of a base, A, with a standard, B, which stops the arm C, controlled by the bar D, when the trigger E is released. The tripping of the trigger is accomplished by the sloping end of D on the slanting end of the upright F. Sides, G, are fastened on the piece F, with their upper ends extending above the bar D, to

which all the working parts are mounted. The upper end of the arm C has a piece, K, to which is attached a

Cannonading a Snow Fort with the Use of a Snowball Thrower

tin can, L, for holding the snowball to be thrown. A set of door springs, M, furnishes the force to throw the snowball.

All the parts are given dimensions, and if cut properly, they will fit together to make the thrower as illustrated.

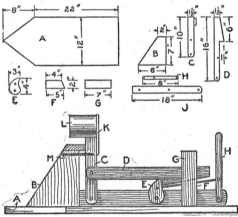

The Dimensioned Parts and the Detail of the Completed Snowball Thrower

prevent the latter from jumping out when it is released by the trigger.

The trigger E is tripped with the handle H, connected to the piece J, on

---

Springs on the Chains of a Porch Swing

Two coil springs of medium strength placed in the chains of a porch swing will make it ride easier and also take up any unpleasant jars and rattles occasioned when a person sits heavily in the swing. If the swing is provided with a four-chain suspension, the springs should be used on the two rear chains to get the best results.—Contributed by E. K. Marshall, Oak Park, Illinois.

Homemade Water Meter

Where it is necessary to measure water in large quantities the meter illustrated will serve the purpose as well

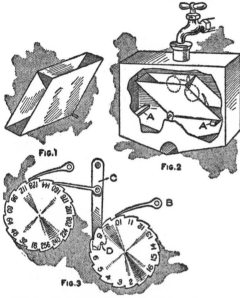

When a Bucket is Filled to the Proper Amount It is Turned Out by the Weight

as an expensive one, and can be made cheaply. The vessel, or bucket, for measuring the water is made diamond-shaped, as shown in Fig. 1, with a partition in the center to make two pockets of a triangular shape, each holding 2 qt., or any amount of sufficient size to take care of the flow of water.

The part forming the pockets is swung on an axis fastened to the lower part, which engages into bearings fastened to the sides of the casing, as shown in Fig. 2. Stops, A, are placed in the casing at the right places for each pocket to spill when exactly 2 qt. of water has run into it. It is obvious that when one pocket is filled, the weight will tip it over and bring the other one up under the flow of water.

The registering device consists of one or more wheels worked with pawls and ratchets, the first wheel being turned a notch at a time by the pawl B, Fig. 3. If each pocket holds 2 qt., the wheel is marked as shown, as each

pocket must discharge to cause the wheel to turn one notch. The second wheel is worked by the lever and pawl C, which is driven with a pin D located in the first wheel. Any number of wheels can be made to turn in a like manner.—Contributed by F. A. Porter, Oderville, Utah.

A Snowball Maker

Snowball making is slow when carried on by hand, and where a thrower is employed in a snow fort it becomes necessary to have a number of assistants in making the snowballs. The time of making these balls can be greatly reduced by the use of the snowball maker shown in the illustration.

The base consists of a board, 24 in. long, 6½ in. wide, and 1 in. thick. A block of wood, A, is hollowed out in the center to make a depression in the shape of a hemisphere, 2½ in. in diameter and 1¼ in. deep. This block is nailed to the base about 1 in. from one end. To make the dimensions come out right, fasten a block, B, 6 in. high, made of one or more pieces, at the other end of the base with its back edge 14½ in. from the center of the hemispherical depression. On top of this block a lever, C, 20 in. long is hinged. Another block, D, is made

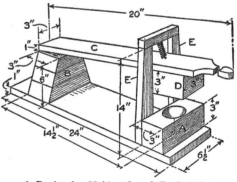

A Device for Making Snowballs Quickly and Perfectly Spherical in Shape

with a hemispherical depression like the block A, and fastened to the under side of the lever, so that the depressions in both blocks will coincide. The lever end is shaped into a handle.

Two uprights, E, are fastened to the back side of the block A as guides for the lever C. A piece is fastened across their tops, and a spring is attached between it and the lever. A curtain-roller spring will be suitable.

In making the balls a bunch of snow is thrown into the lower depression and the lever brought down with considerable force.—Contributed by Abbott W. France, Chester, Pa.

An Inexpensive Bobsled

Any boy who can drive a nail and bore a hole can have a bobsled on short notice. The materials necessary are four good, solid barrel staves; four blocks of wood, 4 in. long, 4 in. wide, and 2 in. thick; two pieces, 12 in. long, 4 in. wide, and 1 in. thick; one piece, 12 in. long, 2 in. wide, and 1¾ in. thick;

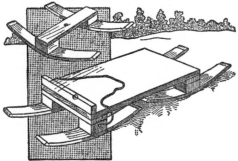

A Bobsled of Simple Construction Using Ordinary Barrel Staves for the Runners

and a good board, 4 ft. long, 12 in. wide, and 1 in. thick.

The crosspieces and knees are made with the blocks and the 1-in. pieces, 12 in. long, as shown; to which the staves are nailed for runners. One of these pieces with the runners is fastened to one end of the board, the other is attached with a bolt in the center. The 1¾ by 2-in. piece, 12 in. long, is fastened across the top of the board at the front end. A rope fastened to the knees of the front runners provides a means of steering the sled.

The sled can be quickly made, and it will serve the purpose well when an expensive one cannot be had.—Contributed by H. J. Blacklidge, San Rafael, Cal.

Motor Made of Candles

A tube of tin, or cardboard, having an inside diameter to receive a candle snugly, is hung on an axle in the center

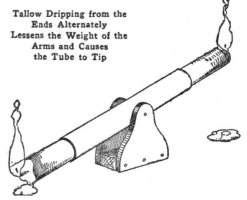

Tallow Dripping from the Ends Alternately Lessens the Weight of the Arms and Causes the Tube to Tip

that turns in bearings made of wood. The construction of the bearings is simple, and they can be made from three pieces of wood as shown. The tube should be well balanced. Pieces of candle are then inserted in the ends, also well balanced. If one is heavier than the other, light it and allow the tallow to run off until it rises; then light the other end. The alternate dripping from the candles will cause the tube to tip back and forth like a walking beam. It will keep going automatically until the candles are entirely consumed.—Contributed by Geo. Jaques, Chicago.

Kettle-Handle Support

The handle of a kettle lying on the kettle rim will become so hot that it cannot be held in the bare hand. To keep the handle fairly cool it must be supported in an upright position. To do this, form a piece of spring wire in the shape

shown, and slip it over the kettle rim. The shape of the extending end will hold the handle upright and away from the heat.

How to Make a Monorail Sled

A monorail sled, having a simple tandem arrangement of the runners, is very easily constructed as follows: The runners are cut from 1-in. plank

An Exhilarating Glide Accompanied by a Buoyant Sense of Freedom Only Obtained in the Monorail Type

of the size and shape given in the sketch, and are shod with strap iron, 1 in. wide and ¼ in. thick. Round iron or half-round iron should not be used, as these are liable to skid. The square, sharp edges of the strap iron prevent this and grip the surface just as a skate.

The top is a board 6 ft. long and 1 in. thick, securely fastened to the runners as follows: Blocks are nailed, or bolted, on either side of the upper edge of the rear runner and the top is fastened to them with screws. The runner is also braced with strap iron, as shown. The same method applies to the front runner, except that only one pair of blocks are used at the center and a thin piece of wood fastened to their tops to serve as the fifth wheel. The hole for the steering post should

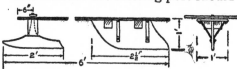

The Construction is Much More Simple Than Making a Double-Runner Bobsled

be 6 in. from the front end and a little larger in diameter than the steering post. The latter should be rounded where it passes through the hole, but square on the upper end to receive the steering bar, which must be tightly fitted in place.

In coasting, the rider lies full length on the board with his hands on the steering bar. This makes the center of gravity so low that there is no necessity for lateral steadying runners, and aside from the exhilarating glide of the ordinary sled, the rider experiences a buoyant sense of freedom and a zest peculiar to the monorail type. Then, too, the steering is effected much more easily. Instead of dragging the feet, a slight turn of the front runner with a corresponding movement of the body is sufficient to change the direction or to restore the balance. This latter is, of course, maintained quite mechanically, as everyone who rides a bicycle well knows.—Contributed by Harry Hardy, Whitby, Ont.

Binding Magazines

To bind magazines for rough service, proceed as follows: Place the magazines carefully one on top of the 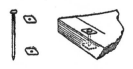 other in order, and space the upper one, near the back edge, for two rivets, marking off three equal distances, or, perhaps, the center space longer than the other two. Make two holes through all the magazines on the marks with an awl, or drill, then drive nails of the right length through them. Use small washers on both ends of the nails under the head and at the point, which is cut off and riveted over. This makes a good, serviceable binding for rough use.—Contributed by Carl W. Lindgreen, Los Angeles, Cal.

A Shellac Cement

As shellac is the basis of almost all cements, a good cement can be made by thickening shellac varnish with dry white lead. The two may be worked together on a piece of glass with a putty knife.

A Blackboard for Children

Take a wide window shade and attach it to a roller as if hanging it to a window. Cut it to about 3 ft. in length, hem the lower edge and insert in the slot in the usual manner. Procure some black slate paint and cover the shade on one side, giving it two coats. Allow sufficient time for the first coat to dry before applying the second coat.

A blackboard of this kind is strong, and if attached to the wall with the shade fixtures, it can be rolled out of the way when not in use.—Contributed by Elizabeth Motz Rossoter, Colorado Springs, Col.

How to Make a Ski Staff

A ski staff will greatly assist progress over level stretches and is an aid to the ski runner in preserving his balance. A homemade staff that is easy to construct is shown in Fig. 1. At the upper end is a narrow leather loop for the wrist; at the extreme lower end a spike is placed for use on icy ground, and just above this spike is a disk, or stop, which, in deep snow, prevents the staff from sinking in too far and gives the necessary leverage for steering, propelling or righting oneself as needed.

The staff is made of a piece of bamboo pole, 1¼ or 1½ in. in diameter, and 4½ ft. long. The leather for the loop can be made from an old strap, shaved down thinner and cut to a width of about ½ in. The stop is a disk of wood, ½ in. thick and 5 in. in diameter. This material should be well-seasoned white pine or spruce and

coated with shellac. A hole is bored through the center of the disk to let it pass upward on the staff about 6 in. Here it is fastened with two

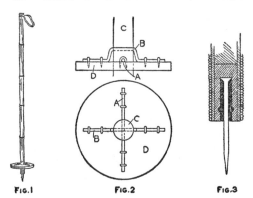

The Staff, being Made of a Bamboo Pole, is Strong as Well as Light

pieces of heavy wire, A and B, Fig. 2. In this diagram, C is the staff, and D, the stop or disk. The wire A passes through the staff below the wire B and at right angles to it, wherefore the wire B must be bent as shown. Both wires are fastened to the stop with staples.

The lower end of the staff, as shown in Fig. 3, is plugged with hard wood, which is bored part way through its center to admit a wire spike. Slight recesses are made in the sides of this hole to anchor the lead which is poured in around the spike. The point of the latter is sharpened and then the bamboo wound with waxed twine, or fine wire, to prevent its splitting.

❧Fine emery cloth, glued to both sides of a piece of bristol board, makes a handy tool for cleaning the platinum points of a vibrator.

A Game Played on the Ice

A novel and interesting winter game for young and old, described as a novelty by a Swedish paper, is played as follows:

Two poles of convenient height are erected on the ice; if skating on a

A Player in Action Ready to Spear a Ring that Hangs on the Line between the Poles

shallow pond they may be driven through the ice and into the ground, but if the water is deep, holes must be bored through the ice and the poles will soon freeze solidly in them. A rope is stretched between the poles at such a height as is suited to the size of the players, or as agreed on to make the game more or less difficult, and on this are strung a number of pieces of board, A, each having a ring of spring steel, B, attached to its lower end. The purpose of the game is to run at good speed between the poles and catch a ring on a spear, each player being entitled to make a certain number of runs, and the winner being the one who can catch the most rings.

The spears may be made of broom handles tapered toward one end, and with a shield made of tin and attached at a suitable distance from the thicker end (Pattern C). The line is fastened at the top of one pole and run through a pulley, D, at the top of the other, thence to a weight or line fastener. Each player should start from the same base line and pass between the poles at such a speed that he will glide at least 100 ft. on the other side of the poles without pushing himself forward by the aid of the skates. Twenty runs are usually allowed each player, or 10 play-

ers may divide into two parties, playing one against the other, etc. An umpire will be needed to see that fair play is maintained and settle any disputes that may arise.

An Electric Display for a Show Window

A novel window display that is very attractive, yet simple in construction and operation, can be made in the following manner: First, make a small watertight chamber, A, as long as the focal length of the lens to be used, and having a glass window, B, at one end, and a small round opening, C, at the other. In this opening is placed a cork through which a glass tube about 2 in. long is inserted. The tube makes a smooth passage for the stream of water flowing out of the box. Water from any source of supply enters the chamber through the tube D, which may be a pipe or hose, whichever is most convenient. The interior is painted a dull black.

A convenient and compact light is placed at the window end of the box. A very good light can be made by placing an electric light with a reflector in a closed box and fastening a biconvex lens, F, in the side facing the window of the water box. When the electric light and the water are turned on, the light is focused at the point where the water is issuing from the box, and follows the course of the stream of water, illuminating it in a pleasing manner.

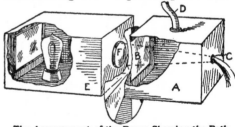

The Arrangement of the Boxes Showing the Path of the Light Rays through the Water

A still better effect can be obtained by passing colored plates between the lens F and the window B. A glass disk with sectors of different colors may be revolved by any source of power, such as a small electric motor or even a waterwheel turned by the flowing water.

Two or three streams of water flowing in different colors make a very pretty display and may be produced by using two or more boxes made up in the same manner. The apparatus should be concealed and nothing but the box end or tube with the flowing water shown.—Contributed by Grant Linton, Whitby, Ont.

Strainer for a Milk Pail

Even though a milker may be careful, small particles of dirt, hairs, etc., will fall into the milk pail. It is true that the milk is strained afterward, but a large percentage of the dirt dissolves and passes through the strainer along with the milk. The best plan to prevent this dirt from falling into the milk is to put a piece of cheesecloth over the pail opening, securing it there by slipping an open wire ring, A, over the rim. The milk will readily pass through the cloth without spattering.—Contributed by W. A. Jaquythe, Richmond, Cal.

Baking Bread in Hot Sand

A driving crew on the river wanted to move camp, but the cook objected as he had started to bake. One of the party suggested using a modified form of the method of baking in vogue more than a century ago, which was to place the dough in the hot earth where a fire had been burning. So, to help the cook out, a barrel was sawed in half and the bread, after being properly protected, was placed in each half barrel and covered with hot sand. Two of the men carried the half barrels on their backs. When the new camp was reached the bread was done.—Contributed by F. B. Ripley, Eau Claire, Wis.

How to Make Small Cams

In making models of machinery or toy machines, cams are very often required. A simple way of making these

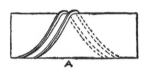

Channels of the Cams Formed with Strips of Brass Soldered to the Drum or Disk

is to lay out the cam plate, or drum, and then bend pieces of brass to the correct shape and solder them in place, whereupon they may be smoothed up with a file or scraper. A cam of this sort on a drum is shown in the sketch at A, and on a faceplate, at B. The method is not quite as accurate as milling, but answers the purpose in most cases.—Contributed by Chas. Hattenberger, Buffalo, N. Y.

Display Holder for Coins

If the luster of coins fresh from the mint is to be preserved, they must be immediately placed so as to be protected against contact with the hands. A good holder that will display both sides of a coin can be made of two pieces of glass, BB, between which is placed a cardboard cut as shown at A. The cardboard should be about the same thickness as the coins. The glass may be framed by using strips

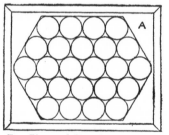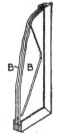

Two Pieces of Glass Inclosing between Them Coins of the Same Size and Thickness

of wood rabbeted to receive the edges of both pieces; or their edges may be bound with passe-partout tape. Even when a frame is used, it is best to bind the edges as this will prevent tarnish from the air. Old negative glass is suitable for making the holder.—Contributed by R. B. Cole, New Haven, Conn.

Holder for Skates while Sharpening

The base of the holder is cut from a board and should be about 3 in. longer than the skate. Two clamps

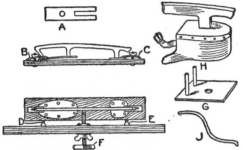

The Holder Provides a Way to Grind a Slight Curve in the Edge of a Skate Blade

are cut as shown at A, from metal of sufficient thickness to hold the skate firmly, then bent to shape and attached to the baseboard with bolts having wing nuts, as shown at B and C.

One edge of the board is provided with two pins, D and E, solidly fastened, which are of sufficient height to bring the center of the blade on a level with the grinder axle. An adjusting screw, F, is provided for the grinder base to adjust the skate blade accurately. The support G is for use on baseboards where skates with strap heels, H, are to be sharpened. The shape of the clamp for this support is shown at J.

When the skate is securely clamped to the base the blade can be easily "hollow ground" or given a slight curve on the edge.—Contributed by C. G. Smith, Brooklyn, N. Y.

A Homemade Direct-View Finder for Cameras

Every hand camera and most of the tripod cameras are equipped with finders of one type or another, and usually one in which the image of the field is reflected upward on a small ground glass—being, in fact, a miniature camera obscura. The later and generally more approved style of finder has a small concave lens conveniently set on the outer edge of the camera. When this direct-vision type of finder is used, the camera is held so that the finder is at the height of the eye, a condition that is particularly desirable. When in a crowd, of course, the professional and many amateurs are familiar with the method of holding the camera inverted over the head and looking up into the finder to determine the range of the field. Even this method is inconvenient, often impractical.

The up-to-date newspaper photographer insists on having his camera equipped with direct finders, as it saves him much trouble and many failures. Anyone with a little ingenuity can change one of the old-type finders into a combination device, either direct or indirect. The sketches are self-explanatory, but it may be said that Fig. 1 represents a box camera with a regulation finder set in one corner of the box. To make it a direct finder, a small brass hinge is used. Cut off part of one wing, leaving a stub just long

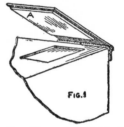

Two Types of Ordinary View Finders and Methods of Converting Them into Direct-View Finders

enough to be attached to the front of the camera directly above the lens of the finder and so as not to interfere with it, and high enough to permit the other wing to be turned down on the ground glass, with space allowed for the thin glass mirror A, that is to be glued to the under side of the long wing. The joint of the hinge should work quite stiffly in order to keep it from jarring out of any position in which it may be set.

If the wing is turned upward at an angle of 45 deg., the finder can be used as a direct-vision instrument when held at the height of the eyes. The image reflected from the small mirror is inverted, but this is no disadvantage to the photographer. The small pocket mirror given out for advertising purposes serves very well for making the reflecting mirror.

The finder shown in Fig. 2 is another very common kind, and one that is readily converted into the direct type by inserting a close-fitting mirror, B, on the inside of the shield to be used as a reflector of the finder image. If the mirror is too thick, it may interfere with the closing of the shield, though in many cases this is not essential, but if it should be necessary to close down the shield in order to fold the camera, it can usually be readjusted to accommodate the mirror.

A Non-Rolling Spool

Bend a piece of wire in the shape shown in the illustration and attach it to a spool of thread. The ends of the wire should clamp the spool slightly and the loop in the wire will keep it from rolling. Place the end of the thread through the loop in the wire and it will not become tangled.—Contributed by J. V. Loeffler, Evansville, Ind.

How to Make a Cartridge Belt

Procure a leather belt, about 2½ in. wide and long enough to reach about the waist, also a piece of leather, 1 in. wide and twice as long as the belt. Attach a buckle to one end of the belt and rivet one end of the narrow piece to the belt near the buckle. Cut two slits in the belt, a distance apart equal to the diameter of the cartridge. Pass the narrow leather piece through one slit and back through the other, thus forming a loop on the belt to receive a cartridge. About ¼ in.

from the first loop form another by cutting two more slits and passing the leather through them as described, and

Two Pieces of Leather of Different Widths Forming a Belt for Holding Cartridges

so on, until the belt has loops along its whole length.

The end of the narrow leather can be riveted to the belt or used in the buckle as desired, the latter way providing an adjustment for cartridges of different sizes.—Contributed by Robert Pound, Lavina, Mont.

Removing Iodine Stains

A good way to chemically remove iodine stains from the hands or linen is to wash the stains in a strong solution of hyposulphite of sodium, known as "hypo," which is procurable at any photographic-supply dealer's or drug store.

There is no danger of using too strong a solution, but the best results are obtained with a mixture of 1 oz. of hypo to 2 oz. of water.

Bed-Cover Fasteners

The arrangement shown in the sketch is easily made and will keep the bed covers in place. The covers are provided with eyelets, either sewed, A, or brass eyelets, B, 6 or 8 in. apart along the edge. A wood strip, C, 3 by 1½ in., is cut as long as the width of the bed and fastened to the frame with wire, bolts, or wedges. Screwhooks, about 1¼ in. long, are turned into the strip so that they will match with the eyelets placed in the covers. Thus the covers will be kept in place

The Hooks Prevent the Covers from Slipping Off the Sleeper and Keep Them Straight on the Bed

when the bed is occupied, and the bed is also easily made up.—Contributed by Warren E. Crane, Cleveland, O.

Collar Fasteners

An excellent fastener to be used on soft collars can be assembled from an ordinary paper fastener and two shoe buttons of the desired color. This device keeps the soft collar in good shape at the front, and serves the purpose just as well as a more expensive collar fastener. The illustration shows how it is used.—Contributed by B. E. Ahlport, Oakland, Cal.

Operating a Bathroom Light Automatically

A device for automatically turning an electric light on and off when entering and leaving the room is illustrated in the sketch. A pull-chain lamp socket is placed upon the wall or ceiling, and is connected to a screw hook in the door by a cord and several rubber bands, as shown.

When the door is opened, the lamp is lit, and when leaving the room the opening of the door again turns it out. The hook should be placed quite close to the edge of the door, to reduce the

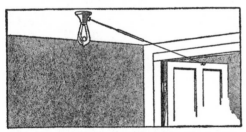

Operating the Electric Lamp Switch or Key by the Opening of the Door

length of the movement, and even then it is too much for the length of the pull required to operate the switch, hence the need of the rubber bands.

The lamp chain pulls out just 1 in., and consequently the lamp is lit when the door is opened part way; and swinging the door farther only stretches the rubber. This is an advantage, however, because the lamp is sure to light regardless of the swing of the door. If no rubber were used, the door would have to open just a certain distance each time.

If the cord is connected to the hook with a loop or a ring, it may be easily disconnected during the day when not needed. A light coil spring may be used in place of the rubbers.—Contributed by C. M. Rogers, Ann Arbor, Michigan.

A Finger-Ring Trick

A coin soldered to some inexpensive ring, or a piece of brass cut from tubing, will make an interesting surprise coin for friends.

The ring when placed on the middle finger with the coin in the palm makes the trick complete. Ask some one if he has ever seen such a coin, or say it is a very old one, as the date is almost worn away. He will try to pick it up, but will find it fast to the finger.—Contributed by Wm. Jenkins, New York City.

Preventing Marks from Basting Threads on Wool

In making up woolen garments it is necessary to press portions of them before removing the basting threads. Sometimes the marks of the basting threads show after the pressing. This can be avoided by using silk thread for basting instead of the usual cotton thread. The silk thread will not leave any marks.—Contributed by L. Alberta Norrell, Gainesville, Ga.

Cranberries will keep fresh for weeks if placed in water in a cool place.

A Bark-Inclosed House Made an
Excellent Home in All Kinds of Weather

A Commodious Shelter Where
Ice Cream and Other
Refreshments were Served

A Nonagon-Shaped Shelter
Provided a Splendid Inclosure
for a Peanut Stand

This Building was Used by a Band
in Giving Concerts Evenings and
on Any Special Holiday

HOUSES MADE OF POLES

BY LOLA A. PINCHON

[In this article descriptions are given of several shelters suitable for a resort, but the reader may select any one of them that answers his needs and build a camp house, or fit up a more substantial one to make living quarters for the whole year.—Editor.]

BEING forced to take the open-air treatment to regain health, a person adopted the plan of building a pole house in the woods, and the scheme was so successful that it was decided to make a resort grounds, to attract crowds during holidays, by which an income could be realized for living expenses. All the pavilions, stands, furniture, and amusement devices were constructed of straight poles cut from young growth of timber with the bark remaining on them. Outside of boards for flooring and roofing material, the entire construction of the buildings and fences consisted of poles.

A level spot was selected and a house built having three rooms. The location was in a grove of young timbers, most of it being straight, and 13 trees were easily found that would make posts 12 ft. long, required for the sides, and two poles 16 ft. long, for the center of the ends, so that they would reach to the ridge. The plot was laid out rectangular and marked for the poles, which were set in the ground for a depth of 4 ft., at distances of 6 ft. apart. This made the house 8 ft. high at the eaves with a square pitch roof; that is, the ridge was 3 ft. high in the center from the plate surfaces for this width of a house. The rule for finding this height is to take one-quarter of the width of

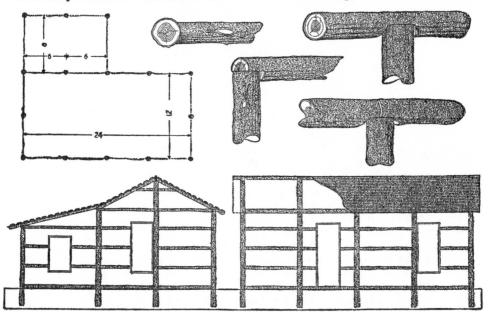

The Frame Construction of the House Made Entirely of Rough Poles, the Verticals being Set in the Ground, Plumbed, and Sighted to Make a Perfect Rectangle of the Desired Proportions

the house for the height in the center from the plate.

The corner poles were carefully lo-

The Steps are Supported on Pairs of Vertical Poles Set in the Ground to Make Different Levels

cated to make the size 12 by 24 ft., with a lean-to 8 by 12 ft., and then plumbed to get them straight vertically. The plates for the sides, consisting of five poles, were selected as straight as possible and their ends and centers hewn down to about one-half their thickness, as shown at A and B, and nailed to the tops of the vertical poles, the connection for center poles being as shown at C.

The next step was to secure the vertical poles with crosspieces between them which were used later for supporting the siding. These poles were cut about 6 ft. long, their ends being cut concave to fit the curve of the upright poles, as shown at D. These were spaced evenly, about 2 ft. apart from center to center, on the sides and ends, as shown in the sketch, and toenailed in place. The doors and window openings were cut in the horizontal poles wherever wanted, and casements set in and nailed. The first row of horizontal poles was placed close to the ground

and used both as support for the lower ends of the siding and to nail the ends of the flooring boards to, which were fastened in the center to poles laid on stones, or, better still, placed on top of short blocks, 5 ft. long, set in the ground. These poles for the floor should be placed not over 2 ft. apart to make the flooring solid.

A lean-to was built by setting three poles at a distance of 8 ft. from one side, beginning at the center and extending to the end of the main building. These poles were about 6 ft. long above the ground. The rafter poles for this part were about 9½ ft. long, notched at both ends for the plates, the ends of the house rafters being sawed off even with the outside of the plate along this edge. The rafter poles for the house were 10 in all, 8 ft. long, and were laid off and cut to fit a ridge made of a board. These poles were notched about 15 in. from their lower ends to fit over the rounding edge of the plate pole, and were then placed directly over each vertical wall pole. They were nailed both to the plate and to the ridge, also further strengthened by a brace made of a piece of board or a

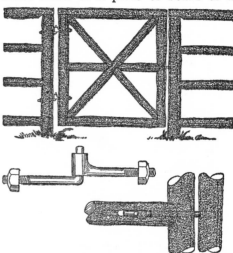

Gate Openings were Made in the Fence Where Necessary, and Gates of Poles Hung in the Ordinary Manner

small pole, placed under the ridge and nailed to both rafters. On top of the rafters boards were placed horizontally, spaced about 1 ft. apart, but this is optional with the builder, as other roofing material can be used. In this instance metal roofing was used, and railing. It is very easy to make ornamental parts, such as shown, on the eave of the porch, by splitting sticks

All Furniture, Together with the Large Lawn Swings, Took on the General Appearance of the Woodland, and As the Pieces were Made Up of the Same Material As the Houses, the Cost Was Only the Labor and a Few Nails

it only required fastening at intervals, and to prevent rusting out, it was well painted on the under side before laying it and coated on the outside when fastened in place. If a more substantial shelter is wanted, it is best to lay the roof solid with boards, then cover it with the regular prepared roofing material.

Some large trees were selected and felled, then cut into 4-ft. lengths and the bark removed, or if desired, the bark removed in 4-ft. lengths, and nailed on the outside of the poles, beginning at the bottom in the same manner as laying shingles, to form the siding of the house. If a more substantial house is wanted, boards can be nailed on the poles, then the bark fastened to the boards; also, the interior can be finished in wall board.

The same general construction is used for the porch, with horizontal poles latticed, as shown, to form the

and nailing them on closely together to make a frieze. Floors are laid on the porch and in the house, and doors hung and window sash fitted in the same manner as in an ordinary house.

A band stand was constructed on sloping ground, and after setting the poles, the floor horizontals were placed about 2 ft. above the ground, on the upper side, and 4 ft. on the lower side. The poles used were about 18 ft. long. Instead of having the horizontals 2 ft. apart, the first was placed 1 ft. above the floor, the next at about one-half the distance from the lower one to the plate at the top, and the space between was ornamented with cross poles, as shown. A balcony or bay was constructed at one end, and a fancy roof was made of poles whose ends rested on a curved pole attached to the vertical pieces. Steps were formed of several straight poles, hewn down on their ends to make a level place to rest on horizontal pieces attached to stakes at the ends. A pair of stakes were used at each end of a step, and these were fastened to a slanting piece at the top, their lower ends being set into the ground. The manner of bracing and crossing with horizontals makes a rigid form of construction, and if choice poles are selected for the step pieces, they will be comparatively level and of sufficient strength to hold up all the load put on them. The roof of this building was made for a sun

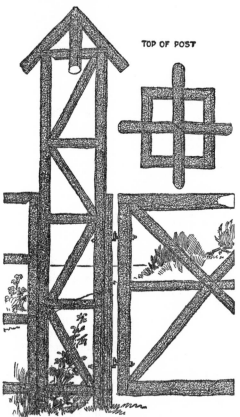

TOP OF POST

The Entrance to the Grounds was Given an Inviting Appearance with Large Posts and Swinging Gates

shade only and consisted of boards nailed closely together on the rafters.

An ice-cream parlor was built on the same plan, but without any board floor; the ground, being level, was used instead. There were five vertical poles used for each end with a space left between the two poles at the center, on both sides, for an entrance. This building was covered with prepared roofing,

so that the things kept for sale could be protected in case of a shower.

A peanut stand was also built without a floor, and to make it with nine sides, nine poles were set in the ground to form a perfect nonagon and joined at their tops with latticed horizontals. Then a rafter was run from the top of each post to the center, and boards were fitted on each pair of rafters over the V-shaped openings. The boards were then covered with prepared roofing. A railing was formed of horizontals set in notches, cut in the posts, and then ornamented in the same manner as for the other buildings.

Fences were constructed about the grounds, made of pole posts with horizontals on top, hewn down and fitted as the plates for the house; and the lower pieces were set in the same as for making the house railing. Gates were made of two vertical pieces, the height of the posts, and two horizontals, then braced with a piece running from the lower corner at the hinge side to the upper opposite corner, the other cross brace being joined to the sides of the former, whereupon two short horizontals were fitted in the center. A blacksmith formed some hinges of rods and strap iron, as shown, and these were fastened in holes bored in the post and the gate vertical. A latch was made by boring a hole through the gate vertical and into the end of the short piece. Then a slot was cut in the side to receive a pin inserted in a shaft made to fit the horizontal hole. A keeper was made in the post by boring a hole to receive the end of the latch.

Large posts were constructed at the entrance to the grounds, and on these double swing gates, made up in the same manner as the small one, were attached. These large posts were built up of four slender poles and were considerably higher than the fence poles. The poles were set in a perfect square, having sides about 18 in. long, and a square top put on by mitering the corners, whereupon four small rafters were fitted on top. The gates were swung on hinges made like those for the small gate.

Among the best and most enjoyed amusement devices on the grounds were the swings. Several of these were built, with and without tables. Four poles, about 20 ft. long, were set in the ground at an angle, and each pair of side poles was joined with two horizontals, about 12 ft. long, spreaders being fastened between the two horizontals to keep the tops of the poles evenly spaced. The distance apart of the poles will depend on the size of the swing and the number of persons to be seated. Each pair of side poles are further strengthened with crossed poles, as shown. If no table is to be used in the swing, the poles may be set closer together, so that the top horizontals will be about 8 ft. long. The platform for the swinging part consists of two poles, 12 ft. long, which are swung on six vertical poles, about 14 ft. long. These poles are attached to the top horizontals with long bolts, or rods, running through both, the bottom being attached in the same manner. Poles are nailed across the platform horizontals at the bottom for a floor, and a table with seats at the ends is formed of poles. The construction is obvious.

A short space between two trees can be made into a seat by fastening two horizontals, one on each tree, with the ends supported by braces. Poles are nailed on the upper surface for a seat.

Other furniture for the house and grounds was made of poles in the manner illustrated. Tables were built for picnickers by setting four or six poles in the ground and making a top of poles or boards. Horizontals were placed across the legs with extending ends, on which seats were made for the tables. Chairs and settees were built in the same manner, poles being used for the entire construction.

An Electric Water Heater

Procure the barrel and cap from a hand bicycle pump and prepare them as follows: Make a tube of paper, about double the thickness of a postal card, to fit snugly in the pump barrel and oil it slightly before slipping it into place. Procure some resistance wire of the proper length and size to heat quickly. The wire can be tested out by coiling it on some nonconducting material, such as an earthen jug or glazed tile, and connecting one end to the current supply and running the other wire of the supply over the coil until it heats properly. Cut the resistance at this point and temporarily coil it to fit into the bottom of the pump barrel, allowing one end to extend up through the space in the center with sufficient length to make a connection to supply wires.

Mix some dental plaster to the consistency of thick cream and, while keeping the wire in the center of the pump barrel, pour in the mixture until it is filled to within 1½ in. of the top. Allow the plaster to set for about a day, then remove it from the barrel and take off the paper roll. The coil of wire at the bottom is now straightened out and wound in a coil over the outside of the plaster core, allowing sufficient end for connecting to the supply wires.

Cut two or three disks of mica to fit snugly in the bottom of the pump barrel, also cut a mica sheet to make a covering tube over the coil on the plaster core and insert the whole into the barrel. The two terminals are connected to the ends of a flexible cord which has a plug attached to the opposite end. Be sure to insulate the ends of the wire where they connect to the flexible cord inside of the pump barrel under the cap. In winding the resist-

An Electric Heating Coil Made of Resistance Wire Placed in a Bicycle-Pump Barrel for Boiling Water

ance wire on the core, be sure that one turn does not touch the other. The heater when connected to a current supply and placed in 1 qt. of water will bring it to a boil quickly.—Contributed by A. H. Waychoff, Lyons, Colo.

Camps

By F. S. CHARLES

A good site, pure water in abundance, and a convenient fuel supply, are the features of a temporary camp that should be given first consideration when starting out to enjoy a vacation in the woods. The site should be high and dry, level enough for the tent and camp fire, and with surrounding ground sloping enough to insure proper drainage. A sufficient fuel supply is an important factor, and a spot should be chosen where great effort is not required to collect it and get it into proper shape for the fire.

When locating near streams of water be careful to select a spot above

Wall Tent

high water mark so the ground will not be overflowed by a sudden rise of the stream. Do not select the site

Lean-To of Boughs

of an old camp, as the surroundings are usually stripped of all fuel, and the grounds are unclean.

Division of Work

Clear the selected spot and lay out the lines for the tent, camp fire, etc.

If the camping party consists of more than two persons, each one should do the part allotted to him, and the work will be speedily accomplished. Remember that discipline brings efficiency, and do not be slack about a camp just because it is pleasure. One of the party should attend to the camp fire and prepare the meals while another secures the fuel and water. The tent can be unpacked and the ground cleared by the other members of the party, and when ready, all should assist in raising the tent, especially if it is a large one.

Tents

An ordinary A or wedge tent is sufficient for one or two campers. Where you do not wish to locate permanently, this tent can be set up and taken down quickly. It should have a ring fastened to the cloth in each peak through which to pass a rope or line to take the place of a ridge pole. Such a tent can be pitched be-

Log Cabin

Fire between Two Logs

Fire Built against a Log

tween two trees or saplings, and, after tying the rope to the trees, it can be tightened with a long forked stick, placed under one end of the rope. If two trees are not conveniently located, then two poles crossed and tied together will make supports for one or both ends, the ridge line running over them and staked to the ground.

On a chilly night, the A tent is quite advantageous. The stakes can be pulled on one side and the cloth doubled to make a lean-to, open on the side away from the wind. A fire can be built in front and the deflected heat on the sleeper will keep him comfortable and warm.

For larger parties, the wall tent with a fly is recommended. These tents can be purchased in various sizes. The fly is an extra covering stretched over the top to make an open air space between the two roofs. It keeps the interior of the tent delightfully cool in hot summer weather and provides a better protection from rain. The fly can be made extra long, to extend over the end of the tent, making a shady retreat which can be used for lounging or a dining place.

Protection from Insects

Where mosquitoes and other insects are numerous, it is well to make a second tent of cheesecloth with binding tape along the top to tie it to the ridge pole of the regular tent. The sides should be made somewhat longer than the regular tent so that there will be plenty of cloth to weight it down at the bottom. This second tent should be made without any opening whatever. The occupant must crawl under the edge to enter. The cheesecloth tent is used inside of the ordinary tent, and when not in use it is pushed aside.

Two camps are illustrated showing the construction of a lean-to for a temporary one-season camp, and a log cabin which makes a permanent place from year to year. (A more elaborate and more expensive camp was described in the May issue of this magazine.) The construction of these camps are very simple. The first is made of poles cut in the woods. A ridge pole is placed between two trees or held in place with poles of sufficient

Forked Sticks Supporting Cooking Utensils

length, set in the ground. Poles are placed on this at an angle of about 45 deg., forming a lean-to that will be entirely open in front when finished. The poles are covered, beginning at the bottom, with pine boughs, laid in layers so as to make a roof that will shed water. A large fire, built a short distance from the open front will make a warm place to sleep, the heat being reflected down the same as described for the A tent.

A Permanent Camp

A good permanent camp is a log cabin. This can be constructed of materials found in the woods. Trees may be felled, cut to length, and notched to join the ends together at each corner so as to leave little or no space between the logs. The roof is constructed of long clapboards, split from blocks of wood. The builder can finish such a camp as elaborately as he chooses, and for this reason the site should be selected with great care.

Camp Fires

There is no better way to make a camp fire than to have a large log or two against which to start a fire with small boughs. Larger sticks can be placed over the logs in such a way as to hold a pot of water or to set a frying pan. Forked sticks can be laid on the log and weighted on the lower end with a stone, using the upper end to hang a cooking vessel over the flames. Two logs placed parallel, with space enough between for the smaller sticks, make one of the best camp cooking arrangements. Two forked sticks, one at each end of the logs, may be set in the ground and a pole placed in the forks lengthwise of the fire. This makes a convenient place for hanging the cooking utensils with bent wires.

Food Supplies

The conditions in various localities make a difference in the camper's appetite and in consequence no special list of food can be recommended, but the amount needed by the average person in a vacation camp for two weeks, is about as follows:

Bacon	15 lb.	Baking Powder	½ lb.
Ham	5 "	Sugar	5 "
Flour	20 "	Beans	4 "
Corn Meal	5 "	Salt	2 "
Rice	5 "	Lard	3 "
	Coffee	3 lb.	

A number of small things must be added to this list, such as pepper, olive oil, sage, nutmeg and vinegar. If the weight is not to be considered, canned goods, preserves, jam and marmalade, also vegetables and dried fruits may be added. Do not forget soap and matches.

Food can be kept cool in a box or a box-like arrangement made of straight sticks over which burlap is hung and kept wet. This is accomplished by setting a pan on top of the box and fixing wicks of cloth over the edges. The wicks will siphon the water out evenly and keep the burlap wet.

A Drinking Tube

When on a walking tour through the woods or country, it might be well to provide a way to procure water for drinking purposes. Take with you several feet of small rubber tubing and a few inches of hollow cane of the size to fit the tube.

In one end insert the cane for a mouthpiece, and allow the other end to reach into the water. Exhaust the air from the tube and the water will rush up to your lips.—Contributed by L. Alberta Norrell, Augusta, Ga.

Washing Photographic Prints

The usual way of washing photographic prints is to place them in a shallow tray in which they will become stuck together in bunches, if they are not often separated. A French magazine suggests that a deep tank be used instead, and that each print be attached to a cork by means of a pin stuck through one corner, the cork thus becoming a float which keeps the print suspended vertically, and at the same time prevents contact with its nearest neighbor.

Camp Furnishings

By CHELSEA CURTIS FRASER

When on a camping trip nothing should be carried but the necessities, and the furnishings should be made up from materials found in the woods. A good spring bed can be made up in the following manner: Cut two stringers from small, straight trees, about 4 in. in diameter, and make them about 6 ft. long. All branches are trimmed off smooth and a trench is dug in the ground for each piece, the trenches being 24 in. apart. Small saplings, about 1 in. in diameter, and as straight as can be found, are cut and trimmed of all branches, and nailed across the stringers for the springs. Knots, bulges, etc., should be turned downward as far as possible. The ends of each piece are flattened as shown at A, Fig. 1, to give it a good seat on the stringers.

A larger sapling is cut, flattened, and nailed at the head of the bed across the stringers, and to it a number of head-stay saplings, B, are nailed. These head-stay pieces are cut about 12 in. long, sharpened on one end and driven a little way into the ground, after which they are nailed to the head crosspiece.

In the absence of an empty mattress tick and pillow cover which can be filled with straw, boughs of fir may be used. These boughs should not be larger than a match and crooked stems should be turned down. Begin at the head of the bed and lay a row of boughs

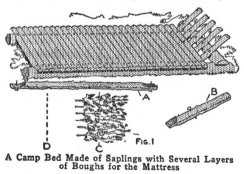

A Camp Bed Made of Saplings with Several Layers of Boughs for the Mattress

with the stems pointing toward the foot. Over this row, and half-lapping it, place another row so that the tops of

the boughs lie on the line C and their stems on the line D. This process is continued until the crosspiece springs are entirely covered, and then another layer is laid in the same manner on top

A Table Made of Packing-Box Material and a Wash Basin Stand of Three Stakes

of these, and so on, until a depth of 6 or 8 in. is obtained. This will make a good substitute for a mattress. A pillow can be made by filling a meal bag with boughs or leaves.

A good and serviceable table can be constructed from a few fence boards, or boards taken from a packing box. The table and chairs are made in one piece, the construction being clearly shown in Fig. 2. The height of the ends should be about 29 in., and the seats about 17 in. from the ground. The other dimensions will be governed by the material at hand and the number of campers.

A wash-basin support can be made of three stakes, cut from saplings and driven in the ground, as shown in Fig. 3. The basin is hung by its rim between the ends of the stakes.

Wherever a suitable tree is handy, a seat can be constructed as shown in Fig. 4. Bore two 1-in. holes, 8 in. apart, in the trunk, 15 in. above the ground, and drive two pins, about 12 in. long, cut from a sapling into them. The extending ends are supported on legs of the same material. The seat is made of a slab with the rounding side down.

A clothes hanger for the tent ridge

pole can be made as shown in Fig. 5. The hanger consists of a piece, 7 in. long, cut from a 2-in. sapling, nails be-

A Seat Against the Trunk of a Tree, and a Clothes Hanger for the Tent Ridge Pole

ing driven into its sides for hooks. The upper end is fitted with a rope which is tied over the ridge pole of the tent.

A Fruit Stemmer

In the berry season the stemmer shown in the sketch is a very handy article for the kitchen. It is made of spring steel and tempered, the length being about 2½ in. The end  used for removing the stem is ground from the outside edge after tempering. A ring large enough to admit the second finger is soldered at a convenient distance from the end on one leg.—Contributed by H. F. Reams, Nashville, Tennessee.

A Homemade Fountain Pen

A very serviceable fountain pen can be made from two 38-72 rifle cartridges and a steel pen. Clean out the cartridges, fit a hardwood plug tightly in

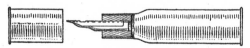

One Cartridge Shell Makes the Fountain Part of the Pen, and the Other the Cap

the end of one shell, and cut it off smooth with the end of the metal. Drill

a $\frac{3}{16}$-in. hole in the center of the wood plug and fit another plug into this hole with sufficient end projecting to be shaped for the length of the steel pen to be used. The shank of the pen and the plug must enter the hole together. One side of the projecting end of the plug should be shaped to fit the inside surface of the pen and then cut off at a point a little farther out than the eye in the pen. On the surface that is to lie against the pen a groove is cut in the plug extending from near the point to the back end where it is to enter the hole in the first plug. The under side of the plug is shaped about as shown.

The other cartridge is cut off at such a point that it will fit on the tapering end of the first one, and is used for a cap. The cartridge being filled with ink and the plug inserted, the ink will flow down the small groove in the feeder plug and supply the pen with ink. Care must be taken that the surface of the smaller plug fits the pen snugly and that the groove is not cut through to the point end. This will keep the ink from flooding, and only that which is used for writing will be able to get through or leak out.—Contributed by Edwin N. Harnish, Ceylon, Canada.

Destroying Caterpillars on Grapevines

The grapes in my back yard were being destroyed by caterpillars which could be found under all the large leaves. The vine was almost dead when I began to cut off all the large leaves and those eaten by the caterpillars, which allowed the sun's rays to reach the grapes. This destroyed all the caterpillars and the light and heat ripened the grapes.—Contributed by Wm. Singer, Rahway, N. J.

¶It will require 1 gal. of ordinary mixed calcimine to cover 270 sq. ft. of plastered surface, 180 sq. ft. of brickwork and 225 sq. ft. of average woodwork.

A Camp Provision Box

While on a camping and canoeing trip recently, I used a device which added a touch of completeness to our outfit and made camp life really enjoyable. This useful device is none other than a provision or "grub" box.

From experience campers know that the first important factor in having a successful trip is compactness of outfit. When undertaking an outing of this kind it is most desirable to have as few bundles to carry as possible, especially if one is going to be on the move part of the time. This device eliminates an unnecessary amount of bundles, thus making the trip easier for the campers, and doubly so if they intend canoeing part of the time; and, apart from its usefulness as a provision container, it affords a general repository for the small articles which mean so much to the camper's welfare.

The box proper may be made of any convenient size, so long as it is not too cumbersome for two people to handle. The dimensions given are for a box I

The Provision Box Ready for Use in Camp, the Cover Turned Back on the Brackets and the Legs Extended

larger box is much to be preferred. A glance at the figures will show the general proportions of the box. It may be possible, in some cases, to secure a strong packing box near the required dimensions, thus doing away with the trouble of constructing it. The distinguishing features of this box are the hinged cover, the folding legs, and the folding brackets. The brackets, upon which the top rests when open, fold in against the back of the box when not in use. The same may be said of the legs. They fold up alongside the box and are held there by spring-brass clips.

On our trips we carry an alcohol stove on which we do all of our cooking. The inner side of the top is covered with a sheet of asbestos, this side being uppermost when the hinged top is opened and resting on the folding brackets. The stove rested on this asbestos, thus making everything safe. The cover is large enough to do all the cooking on, and the box is so high that the cooking can be attended to without stooping

The Brackets for the Cover as Well as Each of the Four Legs Fold Against the Sides of the Box in Such a Manner as to be Out of the Way, Making the Box Easy to Carry and Store Away in a Small Space

used on a canoe trip of several hundred miles; and from experience I know it to be of a suitable size for canoeists. If the camper is going to have a fixed camp and have his luggage hauled, a

over, which is much more pleasant than squatting before a camp fire getting the eyes full of smoke. The legs are hinged to the box in such a manner that all of the weight of the box rests on the legs rather than on the hinges, and are kept from spreading apart by wire turnbuckles. These, being just bolts and wire, may be tucked inside the box when on the move. The

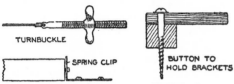

Detail of the Turnbuckle, Button to Hold the Brackets, and the Spring Clip for Holding the Legs on the Side of the Box

top is fitted with unexposed hinges and with a lock to make it a safe place for storing valuables.

In constructing the cover it is well to make it so that it covers the joints of the sides, thus making the box waterproof from the top, if rain should fall on it. A partition can be made in one end to hold odds and ends. A tray could be installed, like the tray in a trunk, to hold knives, forks, spoons, etc., while the perishable supplies are kept underneath the tray. Give the box two coats of lead paint, and shellac the inside.

The wire braces for the legs are made as follows. Procure four machine bolts, about ¼ in. in diameter and 2 in. long—any thread will do—with wing nuts and washers to fit. Saw or file off the heads and drill a small hole in one end of each bolt, large enough to receive a No. 16 galvanized iron wire. Two inches from the bottom of each leg drill a hole to take the bolt loosely. Determine the exact distance between the outside edges of the legs when the box is resting on them. Make the wire braces 1 in. longer than this distance so that the bolts will protrude through the holes in the legs and allow for putting on the nuts and washers. Screwing up on the nuts draws the wire taut, thus holding the legs firm.

The size of the top determines the dimensions of the folding brackets which support it when open. These brackets may be solid blocks of wood, but a lighter and more serviceable bracket is constructed as follows. If the top is 20 in. wide and 30 in. long,

make the brackets 10 by 13 in. Constructing the brackets so that their combined length is 4 in. shorter than the total length of the box, facilitates their folding against the back of the box when not in use. This point is clearly shown in the drawing. Our brackets were made of ½-in. oak, 1½ in. wide, and the joints halved together. They are hinged to the back of the box as shown; and when folded are held in place by a simple catch. The weight of the lid is sufficient to hold the brackets in place when open, but to make sure they will not creep when in use insert a ¼-in. dowel in the end of each so that it protrudes ¼ in. Drill two holes in the top to the depth of ¼ in., so that when the top rests on the brackets, these holes engage with the dowels. In hinging the brackets to the back see that they are high enough to support the lid at right angles to the box.

The box here shown is made of ⅞ in. white pine throughout. The legs are ⅞ by 2½ by 18 in. They are fastened to the box with ordinary strap hinges. When folded up against the box they do not come quite to the top so that the box should be at least 19 in. high for 18-in. legs. About 2 in. from the bottom of the legs drive in a brad so it protrudes ⅛ in. as shown. This brad engages in a hole in the spring-brass clip when folded up as shown in the illustration.

If in a fixed camp, it is a good idea to stand the legs in tomato cans partly full of water. This prevents ants from crawling up the legs into the box, but it necessitates placing the wire braces higher on the legs.

Our box cost us nothing but the hardware, as we knocked some old packing boxes to pieces and planed up enough boards to make the sides. Of course, the builder need not adhere to these dimensions, for he can make the size to suit his requirements, while the finish is a matter of personal taste.

———

❦A blue writing ink is easily made of 1 oz. Prussian blue, 1½ oz. oxalic acid and 1 pt. of soft water. Shake and allow it to stand until dissolved.

Wall Pockets in a Tent

When camping I find a few wall pockets sewed to the tent walls at the back end provide a convenient means to hold the soap, mirror, razor and other small articles liable to be lost. The pockets can be made of the same material as the tent and sewed on as a patch pocket.—Contributed by A. M. Barnes, Atlanta, Ga.

Camp Stoves

The camp stoves illustrated are different forms of the same idea. Both can be taken apart and laid flat for packing. Iron rods, ½ in. in diameter, are used for the legs. They are sharpened at the lower end so that they may be easily driven into the ground. The rods of the one shown in the first illustration are bent in the form of a hook at the upper end, and two pieces of light tire iron, with holes in either end, are hung on these

Camp-Stove Top, Either Solid or Pieced, Supported on Rods at the Corners

hooks. Across these supports are laid other pieces of the tire iron. In the other stove, the rods have a large head and are slipped through holes in the four corners of the piece of heavy sheet iron used for the top. A cotter is slipped through a hole in each rod just below the top, to hold the latter in place.—Contributed by Mrs. Lelia Munsell, Herington, Kansas.

Attractor for Game Fish

A piece of light wood, shaped as shown and with four small screweyes attached, makes a practical attractor for game fish, such as bass, etc., by its action when drawn through the water or carried by the flow of a stream. Hooks are attached to three of the screweyes and the fourth one, on the

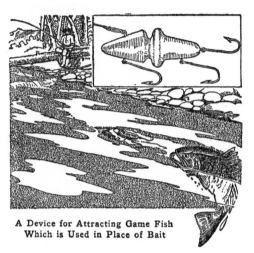

A Device for Attracting Game Fish Which is Used in Place of Bait

sloping surface, is used for the line.—Contributed by Arthur Vogel, Indianapolis, Ind.

Simple Photographic-Print Washer

The ordinary washbowl supplied with a faucet may be easily converted into a washing tray for photographic prints or film negatives. Procure a medicine dropper from a druggist, and attach it to the faucet end with a short piece of rubber tubing. Be sure to procure a dropper that has the point turned at right angles to the body.

The Whirling Motion Set Up by the Forced Stream at an Angle Thoroughly Washes Prints

When the water is turned on it is forced through the small opening in the dropper in such a manner that the water in the bowl is kept in a constant whirling motion. This will keep the prints on the move, which is necessary for a thorough washing.

How to Make an Electric Fishing Signal

A unique electric fishing signal, which may be rigged up on a wharf or pier, and the electric circuit so ar-

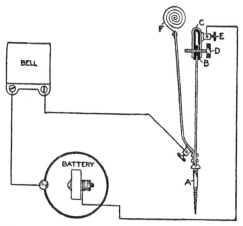

Construction of the Parts to Make the Contact Points and the Electric Connections

ranged as to operate an electric bell or buzzer, located in the fisherman's cottage, or any other convenient place, may be constructed as follows: Obtain two pieces of $\frac{1}{16}$-in. spring brass, one 6 in. long and ¾ in. wide, and the other 7 in. long and ½ in. wide. Mount a 2-in. brass wood screw, A, in one end of the 6-in. piece as shown.

Place over the end of the 6-in. piece a thin sheet of insulating fiber, B, allowing it to extend down on each side about 1 in. Then bend a piece of $\frac{1}{16}$-in. brass, C, over the insulating fiber, allowing it to extend down on each side the same distance as the insulating fiber. Drill a small hole through the lower ends of the U-shaped piece of brass, C, the insulation, B, and the 6-in. piece, while they are all in place. Remove the insulation and the U-shaped brass piece, and tap the holes in the brass for a machine screw, D. Enlarge the hole in the 6-in piece, and provide an insulating bushing for it with an opening of the same diameter as the brass machine screw. Mount a small binding post, E, on one side of the U-shaped piece of brass, and the parts may then be put together and

held in place by means of the brass screw.

Drill two holes in the other end of the 6-in piece, also two holes in one end of the 7-in. piece, and rivet them together with two small rivets. The 7-in. piece should project beyond the end of the 6-in. piece. A piece of thin spring brass should be made into the form of a spiral, F, and fastened to the upper end of the 7-in. piece. Provision should be made for attaching the fishline to the inside end of the brass spiral. A small binding post should be soldered to either the 6-in. or 7-in. piece, at the bottom.

If the device is set up with the head of the brass adjusting screw in the top of the 6-in. piece, pointing in the direction the line to the fishing hook is to run, and if a fish pulls upon the line, the 7-in. piece is pulled over and touches the point of the adjusting screw. If a battery and bell, or buzzer, is connected as shown, the circuit will be completed when the 7-in. piece comes in contact with the adjusting screw, and the bell will ring.

A Chair Swing

A comfortable porch or lawn swing can be easily and quickly made with a chair as a seat, as follows. Procure some rope of sufficient strength to bear

The Ropes are Tied to the Chair so That It will be Held in a Reclining Position

the weight of the person, and fasten one end securely to one of the front legs of the chair and the other end to the same side of the back as shown in the illustration, allowing enough slack to form a right angle. Another piece of rope, of the same length, is then attached to the other side of the chair. The supporting ropes are tied to these ropes and to the joist or holding piece overhead.—Contributed by Wm. A. Robinson, Waynesboro, Pa.

Another Broom Holder

Of the many homemade devices for holding a broom this is one of the simplest, and one that any handy boy can make.

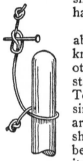

It consists of a string, about 1 ft. long, with a knot at one end and the other tied to a nail or staple driven into the wall. To hang up the broom simply turn the string around the handle as shown, and the broom will be held securely, because its weight will pull the string taut and the knot at the end will prevent the string from running off the handle.—Contributed by Jef De Vries, Antwerp, Belgium.

Squaring Wood Stock

The device shown in the sketch is a great help to the maker of mission furniture as a guide on short cuts. It

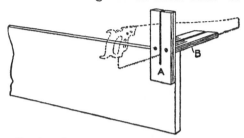

The Saw Teeth Edge can be Run through Both Pieces, the Stock being in the Corner

consists of two pieces of wood, A and B, preferably of oak, fastened together at right angles by two large flat-head screws. The pieces should be placed exactly at right angles.

A cut is then made through both pieces. The cut on B should be exactly at right angles to the surface of piece A. This device can be either clamped on a board or merely held by hand, and will insure a true cut.—Contributed by F. W. Pumphrey, Owensboro, Ky.

A Wind Vane

A novelty in wind vanes is shown in the accompanying sketch. The vane can be made of sheet metal or carved from light wood. The wings are so set on the body as to cause the dragon to rise when the wind strikes them. The dragon is pivoted on a shaft running through its center of gravity, so it will readily turn with the wind. The tail part may also be made to revolve as the propeller of an aeroplane.

The length and size of the shaft will depend on the dimensions of the dragon, and similarly, the location of the weights on the chains will be determined by its size and weight. Upon these circumstances and the varying velocities of the wind will depend how high the dragon

will rise on its shaft, and the height reached by it will thus serve to indicate—in a relative manner only —the velocity of the wind, but it is also possible to arrange the weights at such distances apart that the dragon will rise to A in a 20-mile wind, to B in a 30-mile wind, to C in a 40-mile gale, and so on, with as many weights as desired. This can be done with the aid of an anemometer, if one can be borrowed for some time, or the device may be taken to the nearest weather bureau to be set. — Contributed by H. J. Holden, Ontario, Cal.

❡Never rock a file—push it straight on filing work.

How to Make a Flutter Ring

The flutter ring is for inclosing in an envelope and to surprise the person opening it by the revolving of the

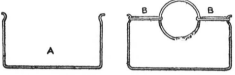

The Shape of the Wire and Manner of Attaching the Rubber Bands to the Ring

ring. The main part is made of a piece of wire, A, bent so that the depth will be about 2 in. and the length 4 in. Procure or make a ring, 2 in. in diameter. The ring should be open like a key ring. Use two rubber bands, BB, in connecting the ring to the wire.

To use it, turn the ring over repeatedly, until the rubber bands are twisted tightly, then lay it flat in a paper folded like a letter. Hand it to someone in this shape or after first putting it into an envelope. When the paper is opened up, the ring will do the rest.—Contributed by D. Andrew McComb, Toledo, O.

A Kitchen Utensil Hanger

Every cook knows how troublesome it is to have several things hanging on one nail. When one of the articles is wanted it is usually at the back, and the others must be removed to secure it. A revolving rack for hanging a can opener, egg beater and cooking spoons, etc., takes up less

The Hook Support Revolves so as to Make Each One Readily Accessible for Hanging Utensils

space than several nails, and places every article within easy reach as well as providing individual hooks for all the pieces.

The rack is easily made of a block of wood, 2½ in. in diameter and 1 in.

thick; an arm, ¾ in. wide, ¼ in. thick and 6 in. long, and a metal bracket. The arm is fastened to the bracket and the bracket to the wall. A screw is turned through a loose-fitting hole bored in the end of the arm and into the disk. Screw hooks are placed around the edge of the dish as hangers.—Contributed by A. R. Moore, Toronto, Can.

Homemade Hinges for Boxes

A very simple form of hinge can be made as shown in the sketch. It is merely a matter of cutting out two pieces of flat steel, A, punching holes in them for screws or nails, and fastening them to the box corners, one on each side. When the box is open, the lid swings back clear and is out

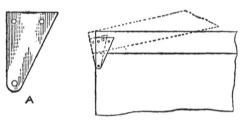

Hinge Parts Made of Sheet Metal and Their Use on a Box Cover

of the way. A hinge of this kind is very strong. For a light box, the parts can be cut from tin.—Contributed by Chas. Homewood, Waterloo, Iowa.

To Remove Odors from Ice Boxes

An easy way to prevent odors in an ice box is to place a can of coke in the box. This will take up all gases and prevent milk from tasting of onions or vegetables which may be kept in the box.

In factories where bad odors are apt to spoil the men's lunches put up in pails or baskets, a box can be constructed to hold these receptacles and a large pail of coke placed in it. Anything placed in this box will remain free from odors, and fresh.—Contributed by Loren Ward, Des Moines, Iowa.

Preventing Window Sash from Freezing to the Sill

When it is cold enough to cause the window sash to freeze fast in the bathroom and bedrooms not having double sash, much discomfort will be experienced and the health may even be menaced. I have discovered a simple method to overcome this difficulty. Lay on the outside sill, close up against the window frame, a thin, narrow strip of wood, on which the window can rest when down. This gives a continual current of fresh air between the sashes at the center, but no unpleasant draft below, and no amount of dripping and freezing will fasten the window sash upon it.—Contributed by Mary Murry, Amherst, Nova Scotia.

A Hanger for the Camp

A garment, or utensil, hanger can be easily made for the camp in the following manner: Procure a long 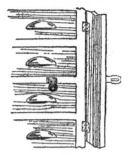 strap, about 1¼ in. wide, and attach hooks made of wire to it. Each hook should be about 4 in. long and of about No. 9 gauge wire. Bend a ring on one end of the wire and stick the other end through a hole punched in the center of the belt. The ring will prevent the wire from passing through the leather, and it should be bent in such a manner that the hook end of the wire will hang downward when the width of the belt is vertical. These hooks are placed about 2 in. apart for the length of the belt, allowing sufficient ends for a buckle and holes. The strap can be buckled around a tree or tent pole.—Contributed by W. C. Loy, Rochester, Ind.

⟨Never stand in a direct line of a swiftly revolving object, such as an emery wheel.

Locking Several Drawers with One Lock

A lock for a number of drawers in a bench or cabinet may be applied with a strip of wood hinged to the cabinet edge so that it will overlap the drawer fronts, as shown. A hasp and staple complete the arrangement for use with a padlock.—Contributed by H. W. Hahn, Chicago.

A Lightning-Calculation Trick

By means of a simple arrangement of numbers, a calculation can be made which will easily puzzle any unsuspecting person. If the two numbers 41,096 and 83 be written out in multiplication form, very few will endeavor to write down the answer directly without first going through the regular work. By placing the 3 in front of the 4 and the 8 back of the 6, the answer is obtained at once, thus: $41,096 \times 83 = 3,410,968$. A larger number which can be treated in the same way is the following: $4,109,589,041,096 \times 83 = 341,095,890,410,968$.

An Adjustable Nutcracker

The advantage of the nutcracker shown in the illustration is that it can be adjusted to various-sized nuts. The handles are similar to those usually found on nutcrackers except that they are slotted at the cracking end to receive a special bar. This  bar is 2 in. long, ½ in. wide, and ⅛ in. thick, with ⅛-in. holes drilled in it at intervals to allow for adjustment. Cotters are used in the holes as pins.

Substitute for a Rubber Stamp

A large number of coupons had to be marked, and having no suitable rubber stamp at hand, I selected a

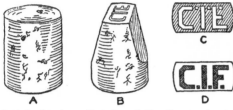

Initials Cut in a Cork Served the Purpose in the Absence of a Rubber Stamp

cork with a smooth end and cut the initials in it. I found that it worked as well, not to say better, than a rubber stamp. An ordinary rubber-stamp pad was used for inking. Angular letters will cut better than curved ones, as the cork quickly dulls the edge of any cutting tool.—Contributed by James M. Kane, Doylestown, Pa.

A Furniture Polish

A good pastelike furniture polish, which is very cheap and keeps indefinitely, can be made as follows: Mix 3 oz. of white wax, 2 oz. of pearlash, commonly known as potassium carbonate, and 6 oz. of water. Heat the mixture until it becomes dissolved, then add 4 oz. of boiled linseed oil and 5 oz. of turpentine. Stir well and pour into cans to cool. Apply with a cloth and rub to a polish. The paste is non-poisonous.

A Hanging Vase

A very neat and attractive hanging corner vase can be made of a colored bottle. The bottom is broken out or cut off as desired and a wire hanger attached as shown. The opening in the neck of the bottle is well corked. Rectangular shaped bottles fitted with hangers can be used on walls.—Contributed by A. D. Tanaka, Jujiya, Kioto, Japan.

Filing Soft Metals

It is well known to mechanics that when lead, tin, soft solder or aluminum are filed, the file is soon filled with the metal and it will not cut. It cannot be cleaned like the wood rasp by dipping it into hot water or pouring boiling water over it, but if the file and the work are kept wet with water, there will be no trouble whatever. Both file and work must be kept thoroughly wet at all times.—Contributed by J. H. Beebee, Rochester, N. Y.

Locking Screws in Door Hinges

When screws once work loose in hinges of doors they will never again hold firmly in the same hole. This trouble can be avoided if the screws are securely locked when they are first put on the door. The sketch shows a

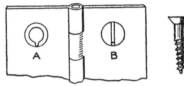

The Screw is Permanently Locked with a Small Nail Driven into the Slot Prepared for It

very successful way to lock the screws. The hole in the hinge for the screw is filed to produce a notch, as shown at A, deep enough to receive a small wire nail or brad, which is driven through the slot in the screw head at one side, as shown at B.

To Remove Grease from Clothing

Equal parts of ether, ammonia and alcohol make a solution that will readily remove grease from clothing. The solution must be kept away from fire, and should be contained in corked bottles as it evaporates quickly, but can be used without danger. It removes grease spots from the finest fabrics and is harmless to the texture.

⊄Jeweler's rouge rubbed well into chamois skin is handy to polish gold and silver articles with.

Paper Glider That Loops the Loop

By C. A. THOMPSON

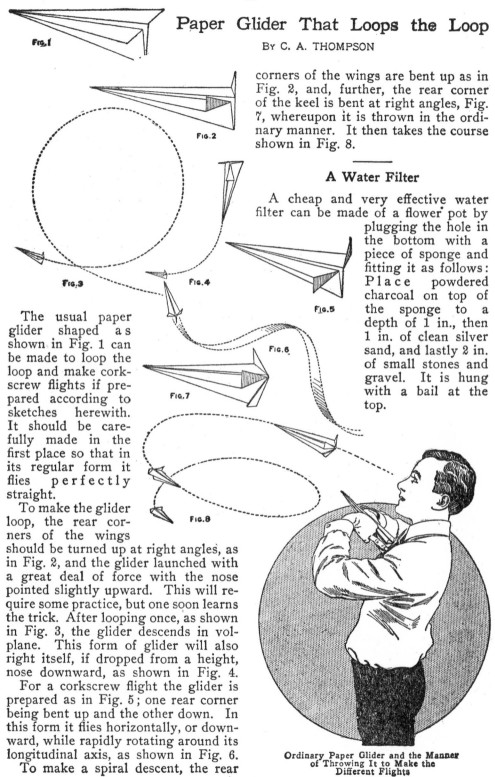

corners of the wings are bent up as in Fig. 2, and, further, the rear corner of the keel is bent at right angles, Fig. 7, whereupon it is thrown in the ordinary manner. It then takes the course shown in Fig. 8.

A Water Filter

A cheap and very effective water filter can be made of a flower pot by plugging the hole in the bottom with a piece of sponge and fitting it as follows: Place powdered charcoal on top of the sponge to a depth of 1 in., then 1 in. of clean silver sand, and lastly 2 in. of small stones and gravel. It is hung with a bail at the top.

The usual paper glider shaped as shown in Fig. 1 can be made to loop the loop and make corkscrew flights if prepared according to sketches herewith. It should be carefully made in the first place so that in its regular form it flies perfectly straight.

To make the glider loop, the rear corners of the wings should be turned up at right angles, as in Fig. 2, and the glider launched with a great deal of force with the nose pointed slightly upward. This will require some practice, but one soon learns the trick. After looping once, as shown in Fig. 3, the glider descends in volplane. This form of glider will also right itself, if dropped from a height, nose downward, as shown in Fig. 4.

For a corkscrew flight the glider is prepared as in Fig. 5; one rear corner being bent up and the other down. In this form it flies horizontally, or downward, while rapidly rotating around its longitudinal axis, as shown in Fig. 6.

To make a spiral descent, the rear

Ordinary Paper Glider and the Manner of Throwing It to Make the Different Flights

A Combination Electrically Operated Door Lock

The illustration shows a very useful application of an ordinary electric door lock in the construction of a combina-

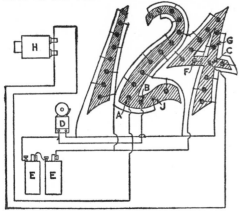

The Brass-Tack Heads Holding the Numerals in Place Constitute the Combination Points

tion lock and alarm to be operated from the outside of the building.

The three numerals, 1, 2, and 4, or any other combination of numbers constituting the house number on a door, are made of some kind of insulating material and fastened in place on a base of insulating fiber, or wood, about ¼ in. thick, by means of ordinary brass-headed tacks, as indicated by the black dots. The tacks will extend through the base a short distance so the electrical connections may be made by soldering wires to them, as shown by the diagram, alternate tacks being connected together with the exception of three; for instance, A, B, and C.

The terminals of the leads that are connected to alternate tacks are in turn connected to the terminals of a circuit composed of an ordinary vibrating bell, D, and battery, E. If any two adjacent tack heads be connected together, except tacks A, B, and C, the bell circuit will be completed and the bell ring, which will serve as an indication that some one is tampering with the circuit. The person knowing the combination, connects the tack heads A and B, and at the same time connects the tack head C with F or G, or any

other tack head that is connected to the plus side of the battery, whereby a circuit will be completed through the lock H and the door is opened. Any metallic substance, such as a knife, key, or finger ring, may be used in making the above indicated connection, and there will be no need of carrying a key for this particular door so long as the combination is known.

The base upon which the numbers are mounted and through which the points of the tacks protrude, should be mounted on a second base that has a recess cut in its surface to accommodate the wires and points of the tacks.

The combination may be made more or less complicated, as desired, by connecting the tacks in different ways, and by using a separate battery for the bell and lock. The circuit leading to the door lock, if there is one already installed, may be used and then no extra circuit is needed.

Such a device has been used on a private-desk drawer with entire satisfaction. The battery was placed in the back end of the drawer, and if it happened to fail, a new one could be connected to the points B and J so that the drawer could be opened and a new battery put in.

Lock for a Fancy Hairpin

To avoid losing a fancy hairpin, bend one leg of the pin as shown in the illustration. The hair caught in the notch

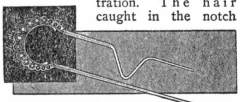

The Bend in the Pin will Hold in the Hair and Prevent the Loss of the Pin

formed by the bend will prevent the pin from dropping out.—Contributed by W. C. Loy, Rochester, Ind.

¶A metal surface polished with oil will keep clean longer than when polished dry.

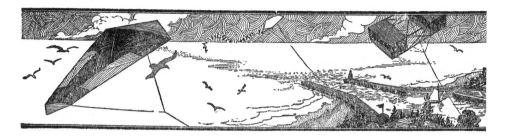

An Aeroplane Kite

By W. A. REICH

After building a number of kites from a recent description in Amateur Mechanics I branched out and constructed the aeroplane kite shown in the illustration, which has excited considerable comment in the neighborhood on account of its appearance and behavior in the air.

The main frame consists of a center-stick, A, 31 in. long, and two cross-sticks, of which one, B, is 31 in. long and the other, C, 15½ in. long. The location of the crosspieces on the centerpiece A is shown in the sketch, the front piece B being 1¾ in. from the end, and the rear piece C, 2¼ in. from the other end. The ends of the sticks have small notches cut to receive a string, D, which is run around the out-

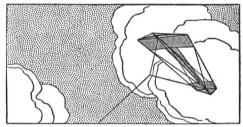

The Kite Being Tailless Rides the Air Waves Like an Aeroplane in a Steady Breeze

side to make the outline of the frame and to brace the parts. Two cross-strings are placed at E and F, 7 in. from either end of the centerpiece A, other brace strings being crossed, as shown at G, and then tied to the cross-string F on both sides, as at H.

The long crosspiece B is curved upward to form a bow, the center of which should be 3¼ in. above the string by which its ends are tied together. The shorter crosspiece is bent and tied in the same manner to make the curve 2½ in., and the centerpiece to curve 1¾ in., both upward. The front and rear parts, between the end and the cross-strings E and F, are covered with yellow tissue paper, which is pasted to the crosspieces and strings. The small wings L are purple tissue paper, 4 in. wide at M and tapering to a point at N.

The bridle string is attached on the centerpiece A at the junction of the crosspieces B and C, and must be adjusted for the size and weight of the kite. The kite is tailless and requires a steady breeze to make it float in the air currents like an aeroplane.

The bridle string and the bending of the sticks must be adjusted until the desired results are obtained. The

bridle string should be tied so that it will about center under the cross-

the air currents properly. The center of gravity will not be the same in the

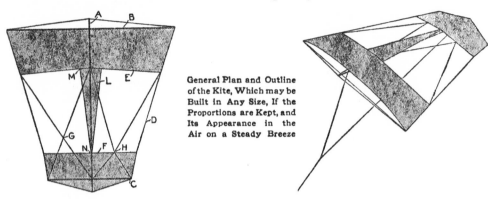

General Plan and Outline of the Kite, Which may be Built in Any Size, If the Proportions are Kept, and Its Appearance in the Air on a Steady Breeze

stick B for the best results, but a slight change from this location may be necessary to make the kite ride

construction of each kite and the string can be located only by trial, after which it is permanently fastened.

Distilling Apparatus for Water

Pure water, free from all foreign substances, is frequently wanted for making up photographic solutions and

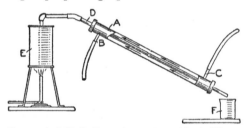

Homemade Still for Removing the Impurities in Water That is Used in Mixing Chemicals

many other purposes. An apparatus for distilling water can be very easily made from galvanized pipe fittings. The outer cooling jacket A is a piece of 1-in. pipe, 2 ft. long, threaded on both ends, and bored and tapped for ½-in. pipe at B and C. A hole is bored and tapped for ½-in. pipe in each of the two caps used on the ends of the pipe A, and a piece of ½-in. pipe, D, 2 ft. 8 in. long, is run through the holes as shown. The joints are soldered to make them water-tight. Two ½-in. nipples, 4 in. long, are screwed in at B and C. The retort, or boiler, E, in which the impure water is boiled may be made of any suitable vessel and

heated with a Bunsen or gas burner. A beaker, or other vessel, F, is placed below the lower end of the small pipe. The cold water from the faucet, which flows into the outer jacket at C and out at B, condenses the steam in the small pipe D, turning it into water which falls into the beaker in large drops. The water is often distilled a second time to remove any impurities which it might still contain.—Contributed by O. E. Tronnes, Evanston, Ill.

Telephone Stand for a Sloping Desk

Having a sloping-top desk and being compelled to use the telephone quite frequently, I devised a support for the telephone so that it might stand level and not fall off. The sides of the stand were cut on the same slope as the desk top, and their under edges were

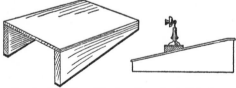

Stand with a Level Surface for a Desk Telephone to be Used on a Sloping Desk Top

provided with rubber strips to prevent slipping.—Contributed by J. M. Kane, Doylestown, Pa.

The Tricks of Camping Out

By STILLMAN TAYLOR

PART I—The Camping Outfit

TO enjoy a vacation in the woods thoroughly, it is essential that the outer be provided with the right kind of an outfit. The inexperienced are likely to carry too much rather than too little to the woods; to include many unnecessary luxuries and overlook the more practical necessities. However,

The Old Hand at the Camping Game Prefers to Cut Poles on the Camping Site and Set Them Up on the Outside for the Camp-Fire Tent

camp life does not mean that one must be uncomfortable, but rather implies plain and simple living close to nature. An adequate shelter from the sun and rain, a comfortable bed, a good cooking kit, and plenty of wholesome food, are the important things to consider. No man or woman requires more, and if unwilling to share the plain fare of the woodsman, the pampered ones should be left at home, for the grouchy, complaining individual makes, of all persons, the very worst of camping companions.

The Choice of a Tent

There are tents and tents, but for average outings in what may be considered a permanent camp, the regulation wall, or army, tent is generally used to make a comfortable shelter. It is a splendid utility tent, with generous floor space and plenty of headroom. For the permanent camp, the wall tent is often provided with a fly, which may be set up as an extra covering for the roof, or extended over the front to make a kind of porch. An extension may also be purchased to serve the same purpose. The 7 by 9-ft. wall tent will shelter two persons comfortably, but when the camp is seldom moved, the 9 by 12-ft. size, with a 3½-ft. wall, will afford more room. The reg-

The Wall Tent may be Erected with the Regular Poles, or, When Ordered with Tapes along the Ridge, It can be Set Up with Outside Tripod or Shear Poles

ulation 8-oz. duck is heavy enough, or the same tent may be obtained in tan or dark green khaki, if preferred. In any case the tent should have a sod cloth, from 6 to 12 in. wide, extending around the bottom and sewed to the tent. An extra piece of canvas or floor cloth is desirable, but this as well as the fly are extras, and while convenient, are by no means necessary. The wall tent may be erected with the regular poles, or it may be ordered with tapes along the ridge and erected by suspending between two trees. The old hand at the camping game rarely uses the shop poles supplied with most tents, but prefers to cut them at the camping site and rig them up on the outside, one slender pole fastened with tapes along the ridge and supported at either end in the crotch formed by setting up two poles, tripod or shear-fashion.

The "Baker" style is a popular tent, giving a large sleeping capacity, yet folding compactly. The 7 by 7-ft. size, with a 2-ft. wall, makes a good comfortable home for two, and will shelter three, or even four, if required. The entire front may be opened to the fire by extending it to form an awning, or it may be thrown back over the ridge to form an open-front lean-to shelter.

The "Dan Beard," or camp-fire, tent is a modification of the Baker style, having a slightly steeper pitch, with a smaller front opening. The dimensions are practically the same as the Baker, and it may be pitched by suspending between two trees, by outside poles, or the regular poles may be used.

For traveling light by canoe or pack, a somewhat lighter and less bulky form of tent than the above styles may be chosen, and the woodsman is likely to select the forester's or ranger types. The ranger is a half tent with a 2-ft. wall and the entire front is open; in fact, this is the same as the Baker tent without the flap. If desired, two half ranger tents with tapes may be purchased and fastened together to form an A, or wedge, tent. This makes a good tent for two on a hike, as each man carries his own half, and is as-

sured a good shelter in case one becomes separated from his companion, and a tight shelter when the two make camp together.

The forester's tent is another good one, giving good floor space and folding up very compactly, a 9 by 9-ft. tent weighing about 5½ lb. when made of standard-weight fabric. It may be had either with or without hood, and is quickly erected by using three small saplings, one along the ridge, running from peak to ground, and one on each side of the opening, to form a crotch to support the ridge pole, shear-fashion. These tents are not provided with sod or floor cloths, although these may be ordered as extras if wanted.

The canoe or "protean" tents are good styles for the camper who travels light and is often on the move. The canoe tent has a circular front, while the protean style is made with a square front, and the wall is attached to the back and along the two sides. Both tents are quickly set up, either with a single inside pole or with two poles set shear-fashion on the outside. A 9 by 9-ft. canoe or protean tent with a 3-ft. wall makes a comfortable home in the open.

Whatever style of tent is chosen, it is well to pay a fair price and obtain a good quality of material and workmanship. The cheaper tents are made of heavier material to render them waterproof, while the better grades are fashioned from light-weight fabric of close weave and treated with a waterproofing process. Many of the cheaper tents will give fair service, but the workmanship is often poor, the grommets are apt to pull out, and the seams rip after a little hard use. All tents should be waterproofed, and each provided with a bag in which to pack it. An ordinary tent may be waterproofed in the following manner: Dissolve ½ lb. of ordinary powdered alum in 4 gal. of hot rain water, and in a separate bucket dissolve ½ lb. of acetate of lead —sugar of lead— in 4 gal. of hot rain water. The acetate of lead is poisonous if taken internally. When thoroughly dissolved, let the solutions

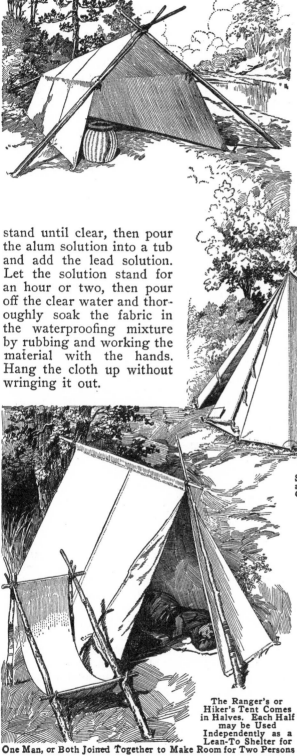

How to Pitch a Tent

It is, of course, possible to pitch a tent almost anywhere, but for the sake of comfort, it is well to select a site with natural drainage. Many campers dig a shallow trench around the tent to prevent water from running in during a heavy rain. This is a good idea for the permanent camp, but is not often necessary if the soil is sandy or porous, or where a sod cloth is used.

It is rarely necessary to

stand until clear, then pour the alum solution into a tub and add the lead solution. Let the solution stand for an hour or two, then pour off the clear water and thoroughly soak the fabric in the waterproofing mixture by rubbing and working the material with the hands. Hang the cloth up without wringing it out.

The Canoe or Protean Tents Are Good Styles for the Camper Who Travels Light and Is Often on the Move, and They can be Quickly Set Up with a Single Inside Pole

carry the regular poles to the camping ground, and they may be omitted excepting when en route to a treeless region. The wall and other large tents may be pitched in several ways. In some places the woodsman cuts a straight ridge pole, about 3 ft. longer than the tent, and two crotched uprights, 1 ft. or more longer than the height of the tent. The ridge pole is passed through the opening in the peak of the tent, or fastened to the outside of the ridge with tapes sewed to the

The Ranger's or Hiker's Tent Comes in Halves. Each Half may be Used Independently as a Lean-To Shelter for One Man, or Both Joined Together to Make Room for Two Persons

cloth. The two upright stakes are then firmly planted in the ground, one at the back and the other in front, and the ridge pole is lifted and dropped into these crotched supports. Set up the four corner guys first to get the tent in shape, then peg down the side guys and slide them taut so that all of them will exert an even pull on the tent. Another good method for setting up the side guys is to drive four crotched stakes, each about 4 ft. long, somewhere near 3 ft. from each corner of the tent, and drop a fairly heavy pole in the rest so formed, then fasten the guy ropes to this pole. When a sod cloth is provided it is turned under on the inside, the floor cloth is spread over it and the camp duffel distributed along the walls of the tent, to hold it down and prevent insects and rain from entering.

To overcome the disadvantage of placing the poles in the center of the entrance, the uprights may be formed by lashing two poles together near the top to make a crotch and spreading the bottoms to form a pair of shears. Poles may be dispensed with entirely, providing the tent is ordered with tapes for attaching a rope to suspend the ridge of the tent between two trees. In a wooded country this manner of setting a tent is generally preferred.

Where a wall tent is used in a more permanent camp, it is a good plan to order a fly, a couple of sizes larger than the tent. This should be set up by using separate poles and rigged some 6 or 8 in. higher than the ridge of the tent, thus affording an air space to temper the heat of the sun and also serving to keep things dry during long, heavy rains.

The Camping Kit

The camping kit, including the few handy articles needed in the woods, as well as the bedding and cooking outfit, may be either elaborate or simple, according to the personal experience and ideas of the camper. In making up a list, it is a good plan to remember that only comparatively few articles are really essential for a comfortable vacation in the wilderness. A comfortable bed must be reckoned one of the chief essentials, and one may choose the de-luxe couch—the air mattress or sleeping pocket—use the ordinary sleeping bag, or court slumber on one of the several other styles of camp beds. The fold-over combination bed, the stretcher bed, or a common bag made of ticking, 6½ ft. long by 2 ft. wide, which is stuffed with browse or leaves, will suffice for the average person. Folding camp cots, chairs, tables, and other so-called camp furniture, have their places in the large, fixed camps, but the woodsman can manage to live comfortably without them. A good pair of warm blankets should be included for each person, providing the sleeping bag is not taken along. The regulation army blankets are a good choice and reasonable in price, or the blankets used at home may be pressed into service.

A good ax is the woodsman's everyday companion, and a good-weight tool, weighing 3 or 4 lb., and a smaller one of 1½ lb. should be carried. When going light, the belt ax will suffice.

The oil lantern is only suited for the fixed camp, since the fuel is difficult to transport unless it is placed in screw-top cans. The "Stonbridge" and other folding candle lanterns are the most convenient for the woods and give sufficient light for camp life.

The aluminum cooking outfits are light in weight, nest compactly, and will stand many years of hard usage, but like other good things, they are somewhat expensive. A good substitute, at half the price, may be obtained in tin and steel, having the good feature of nesting within each other, but, of course, not being quite so light nor so attractive in appearance as the higher-priced outfits. Both the aluminum and steel outfits are put up in canvas carrying bags, and an outfit for two includes a large and a small cooking pot; coffee pot; frying pan with folding or detachable handle; two plates; cups; knives; forks, and spoons. Outfits may be bought for any number of persons,

The Stretcher Bed may be Stuffed with Browse or Leaves, or Suspended from Poles and Stakes to Make a Camp Cot

When Going Light the Belt Ax is Used

A Good, Tempered Knife Should be Worn at the Belt

A Pack Basket with a Waterproof Canvas Lid and Cover, Having Straps to Go over the Shoulders, Is a General Favorite with Woodsmen and Guides

Food Bags with Friction-Top Tins to Fit Them, in Which Lard, Butter, Pork, Ham, and Other Greasy Necessities are Carried

The Cooking Kit may be of Aluminum or Steel, All Nesting within the Largest Pot, and may Include a Folding Baker, or Reflector, with Bread Board in Canvas Bag, a Wood Salt Box, and a Water-Tight Can for Matches

Folding Candle Lanterns are the Most Convenient for the Woods and They Give Sufficient Light for Camp Life and almost all sporting-goods stores carry them. The two-man outfit in heavy aluminum will cost $9 or $10, while the same outfit duplicated in steel is priced at $3.35.

The Camper's Outfit

The personal outfit should include only the most useful articles, and each member of the party should be provided with a dunnage bag of canvas to hold bedding and clothing, and a smaller, or "ditty," bag for keeping together the toilet and other personal belongings which most everyone finds necessary for everyday comfort. A mending kit, containing a few yards of silk, linen, and twist; a length of mending cotton; buttons; a few needles and pins, both safety and the common kinds, should not be overlooked. The veteran usually stows away a bit of wire; a length of strong twine; a few nails and tacks; rivets, etc., for emergency use, and it is surprising to the novice how handy these several odds and ends are found while in camp. A compact tin box will form a convenient place to keep them and will take up little room in the dunnage bag. A medicine case and a first-aid outfit are well worth packing; the smallest cases containing a few of the common remedies will fully meet the camper's needs.

When carrying food by canoe or pack basket, the canoe duffel and provision bags are a great convenience, enabling the outer to carry different foodstuffs in a compact and sanitary manner. Food bags may be had in different sizes, and friction-top tins may be purchased to fit them; and one or more of these liquid-proof containers are desirable for transporting lard, butter, pork, ham, and other greasy necessities. The food bags slip into the larger duffel bags, making a very compact bundle for stowing away in a canoe or pack harness.

Carrying List for the Camp Outfit

For permanent camps, take the wall tent with fly, although the Baker or camp-fire styles are also good. When traveling light by canoe, the canoe or protean tents are recommended. When going very light by pack, use the forester's or ranger's tent. Sod and floor cloths and mosquito netting are optional.

The cooking kit may be of aluminum or steel, all nesting within the largest pot. Include a folding baker, or reflector, with bread board in a canvas bag, a wood salt box, and a water-tight can for matches.

Furniture for the permanent camp consists of a full-sized ax, double-blade or tomahawk style with straight handle, in a protecting case, whetstone and file for keeping the ax in shape. A shovel and saw will be needed when a cabin is built. A canteen may be included, but is not required on most trips. A folding candle lantern is the best for the average trip, but an oil, or acetylene, lantern may be used in a fixed camp. Cots, folding chairs, tables, hangers, etc., are only useful in fixed camps.

A pack basket with a waterproof-canvas lid and cover, having straps to go over the shoulders, is a general favorite with woodsmen and guides. Canvas packs or dunnage bags may be used if preferred. There are two sizes of food bags, one holding 5 lb. and another of 10-lb. capacity, with drawstrings at the top, and these are the best for carrying provisions.

Pack harness, with a tumpline to go across the forehead, is needed when the outfit must be carried on portages, etc. This may be omitted when pack baskets are used. Packing cases of fiber may be used for shipping the outfit to the camping ground, but ordinary trunks, or wood boxes, will answer as well.

The Personal Outfit

An old ordinary suit that is not worn too thin is sufficient. Corduroy is too heavy for the summer and too cold for winter, and canvas is too stiff and noisy for the woods. Cotton khaki is excellent for the summer, and all-wool khaki, or mackinaw, coat and trousers are comfortable for winter. Wool is the best material for undergarments in all seasons. Two sets of garments will be sufficient, as the washing is done at night. Be sure to have the garments large enough to allow for shrinkage. Light-weight cashmere is the best material for socks during summer, and heavier weight for the winter. Three pairs of ordinary-weight and one pair of heavy-weight will be sufficient. A medium-weight gray-flannel overshirt, with breast pockets having button flaps, is the woodsman's choice. On short and light trips one shirt will do. A light-weight, all-wool gray or brown, sweater is a good thing to carry along. It is easily wetted through and a famous brier catcher, yet most woodsmen carry one.

The regulaticn army poncho is more suited to the woods than a rubber coat or oilskins. The larger-size poncho is more bulky to pack, but may be used as a shelter by rigging it up with poles, lean-to fashion. A poncho makes a good ground blanket also.

A medium wide-brimmed hat, in gray or brown, is better than a cap. A gray, or brown, silk handkerchief should be included to wear around the neck to protect it from the sun and cold. Only few novices will carry one, but not so with the regular woodsman. The moccasin is the only suitable footwear for the woods. The "puckaway," with extra sole, is known to most woodsmen. A pair of larrigans—ankle-high moccasins with single sole— are suitable to wear about the camp.

Each member of the party carries his own knapsack, or ditty bag, in which such things as brush and comb, toothbrush, razor, towel, medicines, stationery, etc., are kept. The extra clothing is carried in its own canvas bag.

Each member of the party carries a pair of woolen blankets. Army blankets in tan color are serviceable and inexpensive.

A good, tempered knife should be worn at the belt, preferably one without a hilt and having a blade 5 or 6 in. long.

A small leather pouch containing a few common remedies, such as quinine, laxative, etc.; and a small first-aid out-fit should be included in each camper's personal pack. Also a small leather pouch containing an assortment of needles, darning cotton, buttons, and a length of heavy silk twist is a handy companion.

A few sheets of paper and as many envelopes, a notebook, pencil, and a few postal cards, are usually carried, together with an almanac page of the months covering the intended trip.

The compass is by far the most useful instrument in the woods, but any reliable and inexpensive watch may be carried.

Many woodsmen carry a small hatchet at the belt, and on trips when but the few necessities are carried the belt ax takes the place of the heavier-weight tool. The tomahawk style gives two cutting edges and is therefore the best tool to carry. A leather or other covering case is needed to protect the blades.

A small tin box containing an assortment of rivets; tacks; a bit of string; brass wire; a few nails; a couple of small files; a tool holder with tools; a sheet of sandpaper; a bit of emery cloth, and any other small articles which the sportsman fancies will come in handy, may be carried. It is surprising how often this "what not" is resorted to while in the woods.

The odds and ends of personal belongings, as a jackknife; pipe and tobacco; map of the region visited; length of fishing line and hook; a few loose matches; match box; purse; notebook and pencil; handkerchief, etc., are, of course, carried in the pocket of the coat.

A Camper's Salt-and-Pepper Holder

A camper will find a very clever way to carry salt and pepper by using a piece cut from a joint of bamboo. A piece is selected with the joint in the center, and the ends are stoppered with corks.

A Simple Self-Contained Motor

To say that the subject of this article is the simplest motor in the world is not to overestimate it, for the apparatus 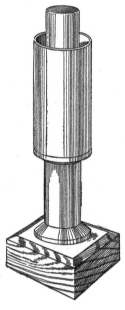 is not only a motor reduced to its essential elements, but combines within itself its o w n source of electric power, all without the use of a s i n g l e piece of wire. The experiment is very interesting a n d instructive and will well repay a careful construction along the lines indicated, even though not in strict accordance w i t h t h e dimensions given.

The first step is to procure a permanent magnet, about ⅜ in. in diameter and 6 in. long. If such a magnet cannot be conveniently secured, a piece of tool steel with flat ends should be hardened by heating it to a dull red and plunging it in water, and then strongly magnetized. This may be readily accomplished by slipping a coil of insulated wire over it through which the current from a storage battery or set of primary cells is passed. If these are not at hand, almost any electrical supply store will magnetize the steel.

A square base block with neatly beveled corners is now in order, which is trimmed up squarely and a hole bored centrally through it to receive the lower end of the magnet. Procure a neat spool and make a hole in it large enough to pass over the magnet. Glue the spool to the base after locating it in the exact center.

The outer and larger cylinder is of copper, or of brass, copperplated on the inside. It is cup-shaped, with a hole in the bottom just large enough to permit the magnet to be pushed through with a close fit, to make a good electrical contact. The magnet may be held in place by having it closely fit the spool and the copper cylinder, and by soldering the heads of a couple of small tacks, or nails, to its under side and driving them into the spool. Coat the magnet with pitch, or paraffin, from the top down, and around its connection with the bottom of the cylinder. The small thimble shown at the top should be of brass or copper, and while one can be easily formed of sheet metal and soldered, it is not improbable that one could be made in seamless form from some small article of commerce. In the exact center of the under side of the top of this thimble, make a good mark with a prickpunch, after which a small steel thumb tack should be filed to a fine needle point and placed, point up, exactly central on the upper end of the magnet, to which it is held with a little wax. The smaller cylinder is simply a piece of sheet zinc bent into a true

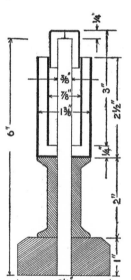

cylinder of such a size that it may be sprung over the lower end of the thimble. This done, it is only necessary to slip the zinc over the end of the magnet until t h e thimble rests on the thumb tack, and then pour some dilute muriatic or sulphuric acid into the outer cylinder, after which the thimble and attached zinc will begin to rotate.

T h e required strength of the acid and the resulting speed will depend upon the nicety of suspension and the trueness of the rotating zinc cylinder. The zinc will have to be changed, but the copper undergoes no deterioration.

The Tricks of Camping Out

By STILLMAN TAYLOR

PART II — Cooking in the Woods

COOKING in the woods requires more of a knack than equipment, and while a camp stove is well enough in a permanent camp, its weight and bulk makes this article of camp furniture unsuited for transportation by canoe. Patent cooking grates are less bulky, but the woodsman can learn to do without them very nicely. However, the important item which few woodsmen care to do without is the folding baker, or reflector. The baker is folded flat and carried in a canvas case, including baking pan and a kneading board. The largest size, with an 18-in. square pan, weighs about 5 lb., and the smallest, with an 8 by 12-in. pan in aluminum, only 2 lb. In use, the reflector is placed with the open side close to the fire, and cooking is accomplished evenly and well in any kind of weather. Bread, fish, game, or meat are easily and perfectly cooked, and the smaller size is amply large for a party of two or three.

The camp fire is one of the charms of the open, and if it is built right and of the best kind of wood, cooking may be done over it as well as over a forest range. Many woodsmen prefer to build a second and smaller fire for cooking, and although I have never found this necessary, excepting in large camps where a considerable quantity of food must be prepared, the camper can suit himself, for experimenting is, after all, a large part of the fun of living in and off the woods.

A satisfactory outdoor cooking range may be fashioned by roughly smoothing the top and bottom sides of two green logs, and placing them about 6 in. apart at one end and about 2 ft. apart at the opposite end. At the wide end a few stones are built up, and across these, hickory, ash, and other sticks of hard wood are placed. The reflector is placed close to the coals at this end, and the fire is built between the logs, the broiling and frying being done at the narrow-end opening. Woods that burn slowly when green should be used for backlogs and end logs; chestnut, red oak, butternut, red maple, and persimmon being best adapted for this purpose.

The hard woods are best for cooking and heating, since they burn more slowly, and give out considerable heat

A Cooking Range Fashioned from Two Green Logs Laid in a V-Shape with a Few Stones Built Up at the Wide End over Which a Fire is Made of Hard-Wood Sticks

and burn down to a body of glowing coals. Soft woods are quick to catch fire, burn rapidly, and make a hot fire, but burn down to dead ashes. Hickory is by far the best firewood of the North, in that it makes a hot fire, is long-burning, and forms a large body of coals that gives an even and intense

A Green Pole Placed in a Forked Stick Provides a Pot Hanger for a Noonday Meal

heat for a considerable length of time. Next to hickory comes chestnut; the basket oaks, ironwood, dogwood, and ash are the woodsman's favorites. Among the woods that are easy to split are the red oak, basket oak, white oak, ash, and white birch. Some few woods split more easily when green than after seasoning, and among them are hickory, dogwood, beech, sugar maple, birch, and elm. The most stubborn woods to split are the elder, blue ash, cherry, sour gum, hemlock, sweet gum, and sycamore. Of the softer woods, the birches make the best fuel; black birch in particular makes a fine camp fire, and it is one of the few woods that burns well when green. The dry bark of the hemlock makes a quick and hot fire, and white birch takes fire quickly even though moist. Driftwood is good to start a fire with, and dry pine knots —the limb stubs of a dead pine tree— are famous kindlers. Green wood will, of course, burn better in winter when the sap is dormant, and trees found on high ground make better fuel than those growing in moist bottom lands. Hard woods are more plentiful on high ground, while the softer woods are found in abundance along the margins of streams.

For cooking the noonday meal a small fire will suffice to boil the pot and furnish the heat sufficient to make a fry. Simply drive a forked stick in the ground and lay a green stick in the fork with the opposite end on the ground with a rock laid on it to keep it down, and hang the pot on the projecting stub left for this purpose. A long stick with projecting stubs, planted in the ground to slant over the fire at an angle, will serve as well. Let the pot hang about 2 ft. from the ground, collect an armful of dry twigs and plenty of larger kindling sticks. Now shave three or four of the larger sticks and leave the shavings on the ends, stand them up beneath the pot, tripod fashion, and place the smaller sticks around them to build a miniature wigwam. While the pot is boiling get a couple of bed chunks, or andirons, 4 or 5 in. in diameter, set and level these on each side of the fire, and put the frying pan on them. When the pot has boiled there will be a nice bed of coals for frying that will not smoke the meal.

When the woodsman makes "one-night stands," he will invariably build the fire and start the kettle boiling while he or a companion stakes the tent, and as soon as the meal is prepared, a pot of water is started boiling for dish washing.

For roasting and baking with the reflector, a rather high fire is needed and a few sticks, a yard or more long, resting upright against a backlog or rock, will throw the heat forward. When glowing coals are wanted one can take them from the camp fire, or split uniform billets of green, or dead, wood about 2 in. thick and pile them in the form of a hollow square, or crib. The fire is built in the center of the crib and more parallel sticks are laid on top until it is a foot or more higher. The crib will act as a chimney, and a roaring fire will result, which upon burning down will give a glowing mass of coals.

Camp cookery implies the preparation of the more simple and nutritious foods, and in making up a list it is well to include only the more staple food-stuffs, whic. are nown to have these qualities. Personal ideas are certain to differ greatly, but the following list may be depended upon and will serve as a guide.

Provision List

This list of material will be sufficient for two persons on an outing of two weeks. Carry in a stout canvas food bag 12 lb. of common wheat flour. The self-raising kind is good, but the common flour is better. It is well to bring a little yellow, or white, corn meal, about 6 lb., to be served as a johnny cake, hot, cold, or fried mush. It is fine for rolling a fish in for frying. Rice is very nutritious, easily digested, and easy to cook. It is good when boiled with raisins. When cold, it can be fried in slices. About 3 lb. will be sufficient. Oatmeal is less sustaining than rice, but it is good for porridge, or sliced when cold and fried. Take along about 3 lb. About 2 lb. of the self-raising buckwheat flour should be taken along, as it is the favorite for flapjacks or griddle cakes. Beans are very nutritious, and about 2 lb. of the common baking kind will be required, to boil or bake with the salt pork. For soups, take 2 lb. of split peas. They can also be served as a vegetable. Salt pork is a stand-by, and 5 lb. of it is provided and carried in friction-top tins or a grease-proof bag. It should be parboiled before adding to the beans or when fried like bacon. The regulation meat of the wilderness is bacon, and 5 lb. of it is carried in a tin or bag. Carry along 3 lb. of lard in a tin or bag, for bread-making and frying. About 3 lb. of butter is carried in a friction-top tin. For making rice puddings, take along 1 lb. of raisins. About 1 lb. of shredded codfish is good for making fish balls. Other small articles, such as ½ lb. of tea; 1 lb. of coffee; 3 lb. of granulated sugar; 1 pt. of molasses; 1 pt. of vinegar; 4 cans of condensed milk; 1 can of milk powder, a good substitute for fresh milk; 1 can egg powder, good for making omelets or can be scrambled; 1 lb. salt; 2 oz. pepper; 1 package each of evaporated potatoes, onions, and fruits, and 3 packages of assorted soup tablets.

This list is by no means complete, but it will suffice for the average person on the average trip, since the occasional addition of a fish or game will help to replenish the stores. When going very light by pack, only the most compact and nutritious foods should be selected, while on short, easy trips the addition of canned goods will supply a greater variety.

Woodcraft

While shooting and fishing and camping out are chapters in the book of woodcraft, the word is generally de-

A Limb Supported at an Angle over the Fire Is Another Means of Hanging the Pot

fined to mean the knack of using the compass, the map, and in making use of the natural signs of the woods when traveling in the wilderness. If the camper keeps to the beaten paths and does not stray far from the frequently used waterways, he needs no compass, and sufficient knowledge of the ways of the woods may be acquired from the previous articles, but if the outer ventures into an unknown region the value of more intimate knowledge increases as the distance to civilization lengthens, because it will enable him to keep traveling in the desired direction and prevent the "insane desire to circle," should one discover he has lost the trail.

The Emergency "Snack" and Kit

The woodsman well knows that it is an easy matter to stray farther from camp than he intended to when starting out, and that it is a common enough occurrence to lose one's bearings and become temporarily lost. To prepare for this possible emergency and spend a comfortable night away from the camp, he carries in his pocket a little packet of useful articles and stows away a tiny package containing a small amount of nutritious food. When leaving camp for a day's hunting and fishing, the usual lunch is, of course, included, but in addition to this, the w o o d s m a n should carry a couple of soup tablets, a piece of summer sausage, and some tea. Wrap this in oiled silk, and pack it in a flat tin box. It will take up very little room in the pocket.

The emergency kit is merely a small leather pouch containing a short fishing line; a few fishing hooks; 1 ft. of surgeon's adhesive plaster; needle and thread; a few safety pins, and a small coil of copper or brass wire. These articles, with the gun and a few spare cartridges, or rod; a belt knife; match safe; compass; map; a little money, pipe, and tobacco, make up the personal outfit without which few woodsmen care to venture far from camp. In addition to the above, I carry a double-edge, light-weight ax, or tomahawk, in a leather sheath at the belt and a tin cup strung to the back of the belt, where it is out of the way and unnoticed until wanted.

The Compass

A small pocket compass affixed to a leather thong should be carried in the breast pocket and fastened to a button of the shirt. An instrument costing $1 will be accurate enough for all purposes. Many of the woodsmen as well as the Indians do not use a compass, but even the expert woodsman gets lost sometimes, and it may happen that the sun is obscured by clouds, thus making it more difficult to read the natural signs of the wilderness. The compass is of little value if a person does not know how to use it. It will not tell in what direction to go, but when the needle is allowed to swing freely on its pivot the blue end always points to the magnetic north. The true north lies a degree or more to either side. In the West, for instance, the needle will be attracted a trifle to the east, while on the Atlantic coast it will swing a trifle to the west of the true north. This magnetic variation need not be taken into account by the woodsman, who may consider it to point to the true north, for absolute accuracy is not required for this purpose. However, I would advise the sportsman to take the precaution of scratching on the back of the case these letters, B = N, meaning blue equals north. If this is done, the novice will be certain to remember and read the compass right no matter how confused he may become on finding that he has lost his way. The watch may be used as a compass on a clear day by pointing the hour hand to the sun, when the point halfway between the hour hand and 12 will be due south.

The compass needle is attracted to iron and steel, therefore keep it away from the gun, hatchet, knife, and other m e t a l articles. Hold the compass level and press the stop, if it has one, so that the needle may swing free. Note some landmark, as a prominent tree, high cliff, or other conspicuous object lying in the direction of travel, and go directly to this object.

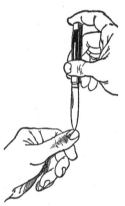

Consult the compass frequently when making a detour, or when the landmark passes out of sight. When this mark is reached, select another farther on and continue the travel, always picking out new marks along the line indicated by the compass. When making camp, consult a map, study it, and so gain a good general idea of the surrounding country; and when leaving camp, take the bearings from the compass. By so doing a person will know in what direction he is traveling, and when the course is changed, keep the general direction in mind. When climbing a hill or making detours, take a mental note of the change in direction and the bearings will not be lost.

Maps

The maps of the U. S. Geological Survey are drawn to a scale of 2 in. to the mile and cost 5 cents each. On the back of each map are printed the

Note Some Landmark, as a Prominent Tree, High Cliff, or Other Conspicuous Object Lying in the Direction of Travel and Go Directly to the Object, and Look for the Old Blaze Marks

symbols showing the character of the land, the contours, roads, and all important rivers and lakes in the district. For convenience, the map should be pasted on a backing of cotton cloth and then cut up into handy sections. Number the sections from left to right and paste a key to the pieces on the back of one of them.

Natural Signs

When traveling through underbrush the woodsman cannot see far ahead, and so lays a true course by noting the position of the sun. For example, here in the northern hemisphere the sun rises just south of east and sets somewhere south of due west. Therefore, if a person is going north, he should keep the sun on the back and to the right shoulder in the morning hours; full on the back at noon, and on the back and over the left shoulder throughout the afternoon.

If the day is cloudy, set the point of a knife blade on the thumb nail, twist it around until the full shadow is cast on the nail, thus indicating the position of the sun.

The direction of the wind is apt to change and for this reason is an unreliable guide, while the so-called signs of the woods, such as the tips of evergreen trees pointing north, bark being thicker on the north side of trees, or moss growing thicker on the north side of the trees, are by no means to be depended upon. There is absolutely nothing in these signs. However, every woodsman is aware that the foliage of trees grows somewhat thicker on the south side, and that the branches are rather shorter and more knotty on the north side, but these and other signs are scarcely infallible, and if they were, few tenderfeet would recognize them.

When traveling by night, look for the Big Dipper or Great Bear, as the two end stars are known as the pointers, pointing to the north star.

Marking the Trail

When traveling over old and blind trails, look for the old blaze marks, and if doubtful about them, make new ones by breaking down the brushes every 15 or 20 ft., the bent part pointing in the direction of travel. If a road is encountered, it is easy to tell if it is a tote or logging road, for tote roads are crooked and wind about the trees and rocks, while the logging road is fairly straight and broad. Of course, tote roads lead nowhere in particular, but all logging roads are sure to come to a fork and lead to water. When breaking a new trail, blaze it by taking a single clip from a tree from the side it is approached, and on the opposite side make two blazes, indicating the way from the camp. If this is done, a person will always know the way back if the trail is crossed from side to side. This is the rule of the wilderness, but is not always observed to the letter, for many woodsmen blaze their trail by clipping the trees as they pass them. Be sure to blaze your own trail correctly, and when you come to a place where two roads or trails fork, set a stick to indicate the right direction.

When a person becomes lost in the woods, as every woodsman is sure to do sometimes, sit down and think it over. Many times a person is nearer camp and companions than it is possible to realize, and if a straight direction is taken, a lumber road or a stream will be found that will give one his bearings. Above all, do not become frightened. If the emergency kit and lunch have not been forgotten, a day and night in the woods alone is not a hardship by any means. Avoid wasting energy by rushing madly about and forgetting to blaze the trail that is being made. Bend the points of the brushes down in the direction of travel. Do not shoot the last cartridge to attract attention, and do not shout until hoarse. Sit down and build a fire of green wood, damp leaves or moss, so that it will smoke. Build a second fire a short distance from the first. This is the recognized signal of the one who is lost. The afternoon may be windy, but the wind is certain to die away at sundown, and the smoke rising from the fires will be visible from a considerable distance. When an Indian gets lost he merely camps on the spot and awaits the next day for picking up the trail.

A Camp Water Bag

While out on a camping trip I devised a way to supply the camp with cool water. A strip of heavy canvas  was cut about 2 ft. long and 1 ft. wide, and the edges were sewed up to make a sack 1 ft. square. In one upper corner a large porcelain knob insulator was sewed in for a mouth piece; the groove around it made a water-tight joint with the cloth. Two metal rings were sewed in the cloth at the top for attaching a strap to carry it. The side and top seams were made as tight as possible.

In use this sack was filled with as cool water as possible and tightly corked. It was then hung in the shade where a breeze would strike it. The water gradually seeped through the cloth and this, in evaporating, kept the contents cool. This sack also came in handy while fishing or on the road.— Contributed by Earl Zander, Three Rivers, Mich.

A Mold for Making Hollow Candy Figures

Those semitransparent candies made up in the shapes of animals which are brought out for the holidays and are so dear to the children have caused many to desire to know how they are made hollow. It is a little trick of the candymaker, which is perfectly clear to mechanics. The candies are cast in metal molds just as babbitt bearings are cast for motor cars. One-half of such a mold is shown in the sketch. A concave recess in the face gives the shape of a horse, dog, or sheep, and another half with a similar recess is laid on and located with two dowels. In use the halves are set on a table resting on the back face A, the hot liquid is poured in at B until the mold is full, then it is allowed to set for a minute, during which the portion in contact with the cold metal hardens, whereupon the mold is turned over and the still liquid center is poured out. This leaves a

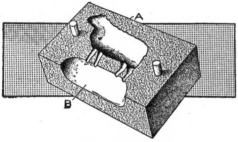

One-Half of a Mold Cut in Metal and Used in Pairs to Make Hollow Candy Objects

glossy surface like candy, a delight to the child, that saves money for the maker and because of its thin walls makes eating easy.

Ornamenting an Old Tree Stump

An old stump remaining after cutting down a large tree in front of a house was made into an ornamental object in the following manner: A cap was made of concrete, reinforced with a square piece of netting. A concrete vase was made and set on the cap. It is only necessary to make a square box from a few scrap boards of the desired size, built up around the stump top. A neat mixture of cement is made and poured in around it, and allowed to set.

Marker for a Hat

A quick and convenient way of marking a hat is to take a visiting card and cut it down in the manner shown in the illustration, then make two small slits in the sweatband of the hat and insert the ends of the card. If the card becomes soiled it can be easily replaced with another. The address can be added if desired.—Contributed by James F. Hatch, Raleigh, N. C.

A Quickly Made Door Latch

A door latch that is efficient as well as simple may be made by bending a piece of iron rod and pointing one end, as shown in the illustration, then securing it to the door with staples; or small rods may be bent in the shape of a staple and the ends threaded for nuts. The door is locked by turning the handle in the position shown by the dotted lines and securing it with a padlock.—Contributed by Claud M. Sessions, Waynesville, Ill.

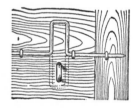

An Electric Lure for Fish

Every good fisherman knows that a light will attract fish. A simple light can be made by taking a pint fruit jar, cutting a 1/4-in. hole in the top of the cover, inserting a piece of gas pipe in the hole and soldering it to the cover. Insulated wires are run through the pipe, and a small electric globe is attached to the ends in the jar. The other ends of the wires are attached to a pocket battery. The jar is placed under water and the light turned on, which attracts the fish.

A Table Box for Campers

By GEO. J. EASTER

A very useful combination packing box and camp table may be made from a coffee or other large box. If a box with a three-ply top is available, it makes a neat appearance, but this is not essential. A box, 14 in. deep, 20 in. wide, and 29 in. long, outside meas-

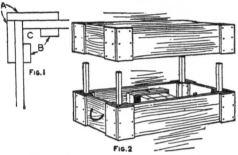

FIG.1

FIG.2

The Strips in the Corners of the Boxes Form Sockets for the Legs

urements, is convenient, as it will slip under the seat of a spring wagon, and is of such a size that a person's knees will pass under it when used as a table.

Saw the box in two on the center line of the narrow way, making two uncovered boxes of the same size and depth. The corners of each box should be well braced on the outside, as shown at A, Fig. 1. The strips B are fastened to the inside of the box to

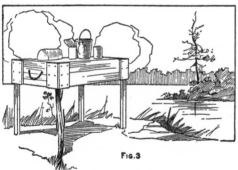

FIG.3

Each Half of the Box Inverted is Used as a Table Top on the Legs

form sockets, C, for the legs. The strips are ½ in. thick, 1¼ in. wide, and as long as the box is deep. Four legs, about 12 in. long and of such size

as to fit in the sockets, are used for holding the boxes together in transit. Rope handles are fastened in the ends of each box, and also a hook and eye, which are used to lock them together.

To pack the boxes place one half open side up, and insert the legs, as shown in Fig. 2. Then fill it and extend the packing to the level of the leg ends; slip the other half of the box on the legs, and fasten the two with the hooks. If properly roped, such a box will be taken as baggage. Canvas, and other articles which will be removed at once upon arrival in camp, rather than provisions, should be packed in this box, so that it can be converted into a table with the least possible work.

To make one table, or two, of the box, remove the packing legs and insert long legs in the sockets of each section. A set of eight legs, 30 in. long, take up very little space, and can be carried diagonally in the bottom of the box. A piece of oilcloth can be wrapped around them and used later as a cover for the table. The legs should fit loosely in the sockets to provide for the swelling in damp weather. Ordinarily they can be wedged to make them rigid. The table is shown in Fig. 3.

Novel Homemade Picture Frames

Pictures can be mounted cheaply and artistically on stiff paper, to make them serviceable for decorating the summer home or camp cottage, without having them placed in a wooden frame. For this purpose a "natural-surface" drawing paper should be used, each sheet being sufficiently large to allow a border all around. With the picture properly centered and marked off, the frame can be fixed. Sets of parallel slots can be cut all around in the border, and a suitable ribbon drawn through so that it is held in place only by the narrow strips of paper. On each of the four corners, a rosette, or similar decoration, can be placed. Ribbons harmonizing with the subject of the picture should be used; for example, a marine scene could have a blue ribbon; fields and brooks, some shade of green; while flowers would probably be best set off by some delicate pink shade.

Another means of decorating the border is to choose some appropriate illustration from a newspaper, or magazine, and carefully trace this outline with carbon paper all around the frame. These figures could then be colored as desired. Profile pictures are best for such work, as they are most easily traced. To be in harmony with the picture, cuts such as captains, sailors, or ships would be suitable for marine

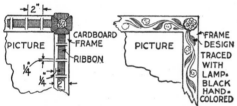

The Edges of Cardboard Extending Out from a Picture Made to Represent a Frame

scenes; soldiers and guns, for war scenes; and trees or flowers, for landscapes.—Contributed by J. B. Murphy, Plainfield, N. J.

Rectangular Opening to Use over Camera View Finder

Ordinary view finders on cameras, having the cut-out in the shape of a Maltese cross, are quite confusing to some camera users. To overcome this difficulty, I cut a piece of celluloid to the shape shown, and in taking a picture, the proper opening is held

The Rectangular Opening Allows Only That Portion of the View to be Seen Which will Show on the Picture

over the view finder so that only the view that will appear in the picture can be seen.—Contributed by E. Everett Buchanan, Elmira, N. Y.

Clipping File Made of Envelopes

Handy pockets for holding notes, or small articles, may be made by anyone from ordinary envelopes. The

The Flaps Hold All the Envelopes Together, Producing a File of Several Compartments

gummed flaps of the envelopes are stuck together after spacing the envelopes to allow a small margin at the end on which the contents of each separate pocket may be written.—Contributed by H. Goodacre, Wolcott, Ind.

Handle for a Drinking Glass

Measure the bottom part of the glass and make a band of copper that will neatly fit it. The ends of the copper can be riveted, but if a neat job is desired, flatten or file the copper ends on a slant, and braze or solder them together.

Attach to the band an upright copper piece a little longer than the glass is high. To this upright piece a bent piece of copper to form a handle is riveted or soldered. The glass is set in the band and the upper end of the vertical piece is bent over the glass edge.—Contributed by William King, Monessen, Pa.

❡A simple and handy pincushion can be made of a large cork fastened to any support or base with a nail or screw.

Combination Camp-Kitchen Cabinet and Table

By J. D. BOYLAN

THE combination camp-kitchen cabinet and table is the result of not being able to take the members of my family on an outing unless they could have some home conveniences on the trip, and perhaps the sketch and description may help solve the same problem for others. The table will accommodate four persons comfortably, and extra compartments may be added if desired. The cabinet, when closed, is strong and compact, and if well made with a snug-fitting cover, is bugproof, and the contents will not be injured greatly, even though drenched by rain or a mishap in a craft.

For coffee, tea, sugar, salt, etc., I used small screwtop glass jars. They are set in pocket shelves at both ends. When closed, one can sit on the box or even walk on it if necessary when

This Outfit Provides Accommodations for Four Persons, and Folds Compactly

in the boat, and if an armful or two of coarse marsh grass is spread over it, the contents will keep quite cool, even when out in the hot sun. When open for use, the metal table top F is supported on metal straps, E, which also act as braces and supports for the table

leaf, G, on each side of the box. This affords plenty of table surface and one can easily get at the contents of the cabinet while cooking or eating. The legs, D, are stored inside of the box when closed for traveling. They are held in place under metal straps when in use, and held at their upper ends by the metal plate and blocks, B and C. The bent metal pieces, A, on the ends of the top, spring over the blocks at B and C, and form the handles.

A Homemade Life Buoy

A serviceable circular life buoy may be made by sewing together rings of canvas, filling the resulting form with ground cork, and waterproofing the covering. Cut two disks of canvas about 30 in. in diameter, and cut out a circular portion from the center of each, about 12 in. in diameter. Sew the pieces together at their edges, leaving a small opening at a point on the outer edge. Fill the cover with cork used in packing grapes, and sew up the opening. Paint the buoy thoroughly, with white lead, and attach hand grips of rope.

Locking Device for Latch Hook on Gate or Door

The troublesome opening of a latch hook on a gate or door, permitting intruders to enter or possibly injuring the door in t h e wind, can be easily overcome by fitting a small catch over t h e hook, as indicated in the sketch. The U-shaped locking device is cut from a piece of tin, and fastened on the screw over which the hook is set. When locked, it is pushed back over the head of the hook, and cannot be easily jarred out of place.

A Vanishing-Cuff Parlor Trick

A trick which is interesting and amusing for the entertainment of the home audience, is performed with a derby hat and a gentleman's cuff. The effect is produced as follows: The performer takes the derby from his head and shows that it is empty. He removes one of his cuffs and drops it into the hat. He tips the hat over so that the spectators can see the inside, and the hat appears empty. He then shakes his arm, and the white cuff reappears, whereupon he places the hat back on his head. The explanation is simple: The white cuff, dropped into the hat, contains a false cuff link, and the inside of the cuff is painted black. A thread holds the cuff in shape until the latter is dropped into the hat, when the thread is broken without the spectators being aware of it. The cuff just fits into the hat, and its ends are deftly snapped beneath the hatband, the hat thus appearing empty. The duplicate cuff is kept on the forearm of the performer, and with a shake, slides into place.—Merritt Hale, Hartford, Conn.

¶A little fresh developer added occasionally to old developing solutions will bring them up in speed and intensity.

Inexpensive Table Lamp Made of Electrical-Fixture Parts

A small table lamp that is light and easily portable, can be made at a cost of less than $1 from electrical-fixture parts, either old or purchased at a supply s t o r e for the job. The base is a bracket, with its brass canopy inverted, as shown. The upright is a ⅛-in. brass pipe, and it is fitted to a standard socket. The shade holder can be made complete from a strip of tin and two wires; or adapted from a commercial shade holder used for candlesticks. Various types of shades, homemade if desired, can be used.

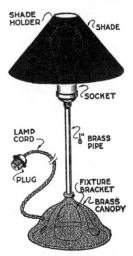

Wire Holders Keep Cabinet Doors Open

Doors of cabinets often have an annoying tendency to swing shut when articles are being removed from the shelves. To overcome this trouble with a kitchen cabinet, I fitted brackets of No. 9 gauge wire into the sides of t h e cabinet, the wire being bent to the shape shown in t h e s k e t c h. When the doors are to be h e l d open, the wires are slid forward from their original position, as indicated by the dotted lines, and set in front of the doors. Before the doors are closed, the wires are quickly snapped back into place. — A. S. Thomas, Amherstburg, Ont., Canada.

"Switchboard" Protects Milker from Cow's Tail

A simple and effective device for guarding a person milking a cow from being hit in the face by the cow's tail

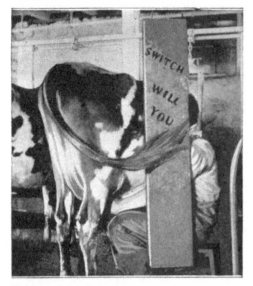

The Legend Put On the "Switchboard" by the Boys Shows How They Value It

is made of a board, about 10 in. wide and 5 ft. long. This is hung by two wire hooks from a long wire running lengthwise of the stable just over the front edge of the gutter. It is moved along with the milker and effectually protects his face while milking. The device was made by a Wisconsin farmer after nearly losing the sight of an eye in being hit by a cow's tail. He tried tying the tails of the cows while milking them, but found by actual test that some cows dropped down as much as 25 per cent in milk production when their tails were tied. The "switchboard" gives the cows the necessary freedom.—D. S. B., Wisconsin Live Stock Breeders' Association.

Reflected-Light Illumination with Homemade Arrangement

"Friend wife" does not complain any longer because of poor light over the kitchen stove. The windows in the kitchen were so disposed that the light was partly shut off from the stove by the person standing before it. I solved the difficulty in this way: A small window was cut directly back of the stove, in a partition between the kitchen and an adjoining storeroom, locating it just a few inches above the top of the stove. A mirror was placed, after some experimenting, so that the light from an outside window in the storeroom was reflected through the small window in the partition and onto the top of the stove. Plenty of light was thus afforded. Various adaptations of this arrangement may be worked out.—F. E. Brimmer, Dalton, N. Y.

Bedroom Shade and Curtains Arranged for Thorough Ventilation

Curtains, shades, and similar fixtures, often interfere with the proper ventilation of sleeping rooms. By arranging these features as shown in the sketch, the ventilation is not interfered with, and the shades and curtains give the same service as with the usual arrangement. The curtains are hung singly on hinged bars, which may be homemade or those used as towel bars. Details of the supports, at A and B, are shown in the sketch. Two pairs of fixtures are provided for the shade, permitting it to be lowered at night, with free circulation of the air at the top and bottom. The shade is quickly raised, and the

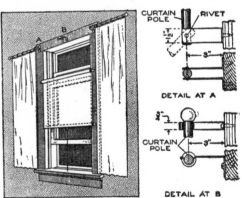

This Arrangement of Curtains and Shade Permits Thorough Ventilation in the Sleeping Room

curtains swung into their closed position.—J. E. McCoy, Philadelphia, Pa.

Coffee Grinder Repaired with Rubber Faucet Plug

A rubber piece that held the glass container on a wall coffee grinder wore out, so that the glass would not stay in, putting the grinder out of commission. The piece worn out was a thick rubber washer, tapered at one end to form a hollow in the other, when in place. I cut a section from the thick end of a standard rubber faucet plug, and shaped it to the form of the desired washer. By removing the old rivet carefully, I was able to use it in fixing the new washer into place, and the mill was soon grinding merrily, as of old.—M. T. C., Chicago, Ill.

Coal Hod Made from Iron Pipe

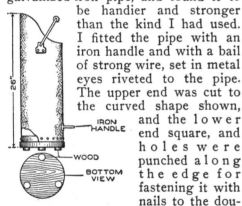

When my coal hod became worn out, I made one of a length of 8-in. galvanized-iron pipe, and found it to be handier and stronger than the kind I had used. I fitted the pipe with an iron handle and with a bail of strong wire, set in metal eyes riveted to the pipe. The upper end was cut to the curved shape shown, and the lower end square, and holes were punched along the edge for fastening it with nails to the double bottom of wood. Three wooden pads were set under the bottom, so that the hod stands easily on a spot that is not quite level.—T. E. Roberts, Toronto, Canada.

Frayed Shoe Laces Repaired with Pitch or Wax

When the tag or end fastening comes off of a shoe lace, take a little black sealing wax, and press it carefully around the end of the lace and shape it to a point. This will last a long time, and does away with the annoyance of frayed lace ends.

An Enameled Armchair Made of Wooden Strips

An armchair suitable for a dressing table was made by a handy woman from pine strips. The photograph

The Simple Construction of This Neat Armchair Makes It an Attractive Job for the Amateur Craftsman

shows the simple and pleasing lines of the construction. Aside from the board seat, only three sizes of wood are used, 2 by 2 in., 1 by 2 in., and ½ by 2 in. The pieces are fastened with screws, round-head brass ones being used at exposed points. The seat is wider from side to side than from front to back. Two coats of white paint and one of white enamel give a good finish.

The dimensions may be varied to suit individual needs. Sizes suggested are: back, 32 in. high and 24 in. wide; side, 26 in. to top of arm and 19 in. wide; seat, 17 in. from floor, 18 in. from front to back, and 20 in. wide between the front supports. The stock is all planed up square to dimensions, and sandpapered smooth. The ends should be cut squarely in a miter box, with a fine-toothed saw, and then sandpapered smooth, taking care not to round the ends.—A. May Holaday, Chico, Calif.

A Curling-Iron Heater

Heating of curling irons is a not uncommon source of fires, and to minimize this danger, an electrical heating

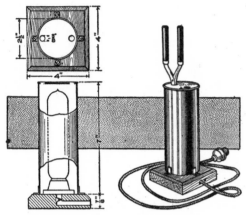

This Efficient Electric Heater for Curling Irons can be Made Quickly and Is Safer to Use than an Open Flame

device is valuable. In the arrangement shown, a long candle-shaped incandescent lamp is mounted in a suitable can, or metal tube, to form the body of the heater. This should be bright, both inside and out, and preferably nickeled. Two irons can be heated by setting them in the holes in the top. Four small brackets, soldered, as shown, around the edge of the can, hold it to the base. A porcelain lamp receptacle is used, mounted on a base block. Stain, fill, and finish the wood as desired. In assembling the parts, screw the receptacle to the base, and connect the flexible cord through a suitable hole. Screw in the lamp, and fasten down the body.—John D. Adams, Phoenix, Arizona.

A Stepmother for Incubator Chicks

The best imitation mother hen for incubator chicks that we have found, is built by attaching rag strings to the bottom of an inverted cracker, or similar, box, which accommodates 2 doz. chicks. The rag pieces are torn 1 in. wide, from coarse cloth or gunny sacking, and their loose ends just touch the floor. They are placed close together.

An inlet to the mother box is cut in the edge of it. The chicks huddle in among the string rags, keeping safe and warm, whereas without such an arrangement, they may crowd together too closely, and some of them be smothered.—J. Cecil Alter, Salt Lake City, Utah.

A Cardboard Writing and Drawing Pad

Where neatness is desired in the writing of themes, manuscripts, etc., especially if sketches are set into the copy, a homemade writing and drawing pad is useful. The one shown in the sketch can be made easily, and is a handy device for school children as well as older persons. The paper is placed under the guide strips, as shown, and is moved along under the sliding straightedge as desired. Drawing instruments can be used handily along the straightedge, as indicated. The pad is built up as detailed in the sectional views. The general dimensions can be varied. The bottom piece is $\frac{1}{8}$-in. cloth board. The second layer is $\frac{1}{32}$ in. thick and of cardboard; the next is $\frac{1}{16}$-in. cardboard, and the upper layer is $\frac{1}{8}$-in. cloth board, similar to the straightedge. The latter rests on the projecting guides for the paper, and

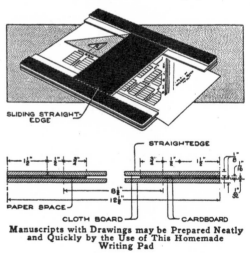

Manuscripts with Drawings may be Prepared Neatly and Quickly by the Use of This Homemade Writing Pad

is set between the shaded sidepieces.—R. S. Edwards, Walla Walla, Wash.

Homemade Shoulder-Pack Tent

By J. D. BOYLAN

AFTER sleeping under various kinds of canvas coverings and not finding any of them entirely to my liking, I made the tent shown in the illustration, which proved quite satisfactory. It is of light weight, easily set up or taken down, and when buttoned closely is practically rain, wind, and bug-proof. The cost of materials necessary for making it is comparatively slight. I use it not only as a sleeping tent but also as a carry-all in packing camping equipment. The canvas is supported by frames made of pliable branches cut in the woods.

The layout for the canvas is shown in the detailed drawings. The sections for the ends are made of three pieces, one for the ground and two, divided vertically, for the end covering. The ground section of the main portion of the tent and the covering are made in one piece, 6 ft. wide, joined at the middle, as shown. The adjoining edges A are sewed together and the

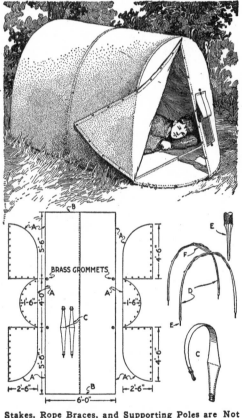

Stakes, Rope Braces, and Supporting Poles are Not Required for This Shoulder-Pack Tent, the Supports being Cut at the Camp

edges B, which are set at the ridge of the tent, are sewed, after the other pieces are joined. Brass grommets are fitted in the canvas, as indicated, and the points of the supporting frames pass through them in driving the supports into the ground. The shoulder straps C are placed so that they are in position when the tent is folded, and rolled into a pack. Other equipment may be placed inside of it. The tent supports D are pointed at the ends E, and are twisted together at the top. The ridge pole F steadies them and holds the canvas at the middle.

To set up the tent, lay the canvas flat on the ground and place the supports, twisted together, through the grommets. Spring them into the ends of the canvas, and insert the ridge pole by springing it between the supports. The canvas is 8-oz. duck, and the fastenings used are snap buttons; buttonholes, buckles, or harness snaps may also be used.

Kitchen for Hikers

By PRESTON HELLER

WITH a view to provide all the needs of a commissary department for 36 boys for a period of four days, either on a hike or in a permanent camp, the kitchen illustrated was constructed. As it is placed on two wheels, which are removed when the kitchen is in use, it can be moved from one day's camp to another by attaching it to the rear of a horse-drawn wagon by means of a shaft. When the wheels are removed the entire outfit rests on legs, which are swung down from the bottom. The sides and one end are opened by swinging one half up and resting it on the top, while the other half swings down to a horizontal position where it is used as a work board, making all parts easily accessible.

The outside dimensions of the kitchen, when closed and in the form of a large box on wheels, are 5 ft. 3 in. long, 3 ft. wide, and 2½ ft. high. The main feature of this entire kitchen is its compactness. At the front, and extending about 1 ft. back, is a kitchen cabinet where the plates, sugar, salt, flour, etc., are kept in separate compartments. Here also are found the necessary cooking utensils, such as bread knives, butcher knives, cleaver, cooking spoons, pancake turner, sieves, large forks, lemon squeezer, etc.; and on the shelves of galvanized iron small boxes and packages of baking powder, cocoa, etc., are placed. This entire compartment, as well as all others where food is handled and prepared, is lined with No. 28 gauge galvanized iron which makes sanitation a feature also.

Upon passing around to one side there can be seen a large three-shelved oven, 21 in. wide, which is heated by a gasoline burner. Between the burner and the bottom of the oven are located coils of pipe for heating water, and these coils are connected with a tank of 7-gal. capacity, located just above the oven. An air valve and glass gauge are attached to the tank.

The next compartment to the rear is a large storage space, extending all the way through the kitchen, and a 2½-gal. forged-copper gasoline tank occupies a shelf in the upper portion of this space. At the rear end along this side are located nickelplated faucets which are connected with the hot-water tank mentioned; a 7-gal., white-enameled milk tank above; an 18-gal. cold-water tank, and an ice-water tank, used when distilled-water ice can be secured. These faucets all drain into a small sink, which, in turn, drains off through an ordinary sink drain to a hole dug in the ground beneath it. Practically the entire rear end of the kitchen is occupied by the large water tanks, ice box, and milk tanks, with the exception of a small space at the bottom where the silverware is kept in a drawer.

On the other side, and to the rear, two compartments above and below the large water tank form excellent storage space for ham, bacon, sausage, preserves, butter, etc., which need to be kept in a cool place. Next in line is the other end of the large storage place which extends through from the other side. Pans, pails, canned goods, larger packages, etc., are kept in this space.

Immediately to the rear of the kitchen cabinet, on this side, are located compartment shelves where the tin cups are kept, and adjoining this is found a three-compartment steam cooker. By having the cups and plates near this steam cooker, which is also heated by a gas burner, there is less danger from rust, as they are kept thoroughly dried. Wherever there is a gasoline burner the compartment in which it is located is not only lined with galvanized iron, but asbestos in sheets is placed on the inner side, so that the heat will not ignite the interior packing or the woodwork. The tanks are accessible from the top of the kitchen for filling and cleaning, and are packed with ground cork.

The kitchen has shown its efficiency by giving satisfactory service in camps of many members.

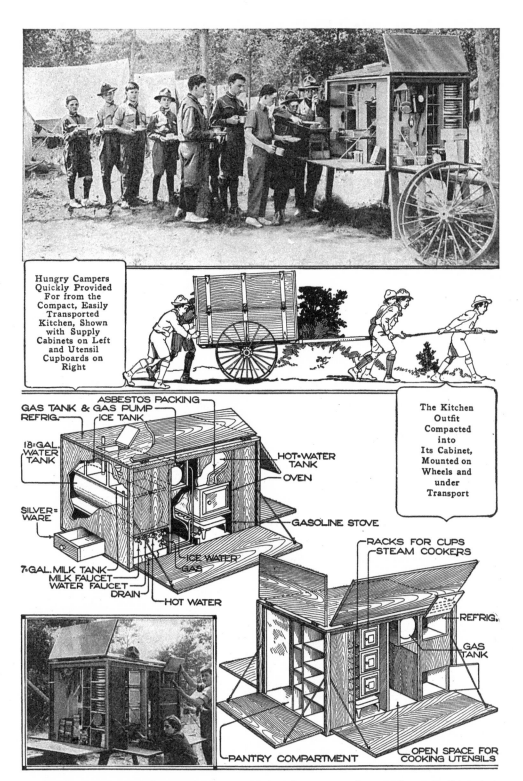

Hungry Campers Quickly Provided For from the Compact, Easily Transported Kitchen, Shown with Supply Cabinets on Left and Utensil Cupboards on Right

The Kitchen Outfit Compacted into Its Cabinet, Mounted on Wheels and under Transport

GAS TANK & GAS PUMP
REFRIG.
ASBESTOS PACKING
ICE TANK
18-GAL. WATER TANK
HOT-WATER TANK
OVEN
SILVER= WARE
GASOLINE STOVE
7-GAL. MILK TANK
MILK FAUCET
WATER FAUCET
DRAIN
ICE WATER
GAS
HOT WATER

RACKS FOR CUPS
STEAM COOKERS
REFRIG.
GAS TANK
PANTRY COMPARTMENT
OPEN SPACE FOR COOKING UTENSILS

The Portable Kitchen Outfit Opened, Exposing the Various Compartments Arranged So as to Be Convenient: Above, Stove and Cooking Compartments; Below, Pantry Compartment and Space for Utensils

Bird House Made of Kegs

Two ordinary nail kegs, or other small kegs, will make a good bird house. They should be mounted on a square post with braces of light wood, as shown. The openings for the entrance can be cut in the ends or sides, as desired. If cut in the sides, be sure to make the hole between two staves.

A Drinking-Glass Holder

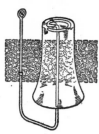

To make a glass holder quickly, shape a wire as shown in the sketch. The wire should be of sufficient size to hold the glass firmly. It is fastened to the wall, or support, with a screw and a staple. — Contributed by Edwin P. Stott, Chicago, Ill.

Needle Threader for a Sewing Machine

The threader consists of two brass pieces riveted together so that they will make an extension conical hole for the thread to enter freely. The length of the pieces should be such that when the upper end of the threader is placed against the needle-holder end, the conical hole will coincide with the hole in the needle.

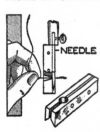

A piece of thin sheet metal is placed between the pieces above the conical hole to make an opening below for the thread to pass through when the threader is removed. The opposite edges of the brass pieces from the large part of the conical hole are filed to a bevel so that when they are riveted together they will form a V-shaped groove to center the needle.—Contributed by Harriet M. Kerbaugh, Allentown, Pa.

Winding Coiled Springs

When a helical spring is needed badly, one can be made up quickly by winding the wire around in the threads of an ordinary bolt. An accurate spring can be formed, and the pitch between each coil will be exact for the entire length.

In removing the spring from the bolt, grasp the coil in one hand and turn the bolt with the other.

Revolving-Wheel Ruling Pen

A ruling pen that will do neat work and not leave any ink on the ruler, and which with its small ink fount draws many lines at one filling, can be made from an old discarded revolving-wheel glass cutter. A ⅛-in.

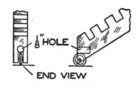

hole is drilled in the body, or handle part, just above the wheel. The hole is filled with a piece of felt—a piece cut from an old felt hat will do—rolled to fit snugly, so that it will bear lightly against the wheel. The felt is soaked with the ink to be used, and the pen is ready for use.—Contributed by Bert Fish, Rochester, N. Y.

⟨An excellent method of closing a crack in a wall before papering is to paste a thin strip of linen over the crack. This not only makes a smooth job, but prevents the paper from tearing, should the crack widen.

A Portable Folding Boat

By STANLEY L. SWIFT

A BOAT that is inexpensive, easily made, and readily transported is shown in the illustration. Since the bow section folds inside of the stern portion, it is important that the dimensions be followed closely. The material used is ⅞-in. throughout.

Make a full-size diagram of the plan to determine the exact sizes of the pieces. Brass screws are best for fastening this type of work, but copper-plated nails may be used. Tongued-and-grooved stock is best for the bottom. The joints should not be driven together too firmly, to allow for expansion, and all joints in the boat should be packed with red lead or pitch.

The adjoining ends of the sections should be made at the same time, to insure a satisfactory fit when joined. Braces are fixed into the corners.

Metal straps hold the sections together at the bottom of the hinged joints. These should be fitted so that there is little possibility of their becoming loosened accidentally. The front end of each strip is pivoted in a hole, and the other end is slotted vertically on the lower edge. Their bolts are set firmly into the side of the boat, being held with nuts on both sides of the wood. A wing nut, prevented from coming off by riveting the end of the bolt, holds the slotted end. Sockets for the oars may be cut into hardwood pieces fastened to the gunwales. The construction of the seats is shown in the small sketch at the left.

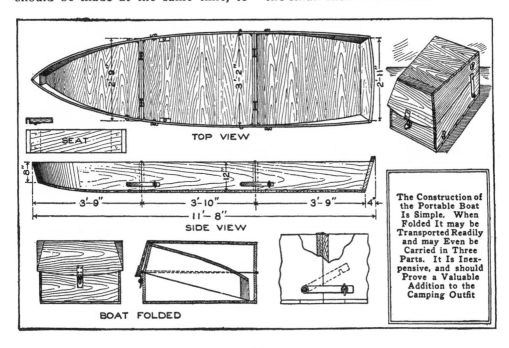

SEAT

TOP VIEW

8"

3'-9"   3'-10"   3'-9"   4"

11'-8"

SIDE VIEW

BOAT FOLDED

The Construction of the Portable Boat Is Simple. When Folded It may be Transported Readily and may Even be Carried in Three Parts. It Is Inexpensive, and should Prove a Valuable Addition to the Camping Outfit

Nontangling Pasture Stake

An old carriage wheel and axle were used to good advantage in the making of the pasture stake shown in the

The Wheel Revolves on the Stake, Preventing the Rope from Tangling

sketch. When the animal tethered to the wheel walks around the stake, it is practically impossible for the rope to become tangled, as the wheel revolves on the axle.—Contributed by W. F. Quackenbush, New York, N. Y.

Inkstand Made of a Sheet of Paper

Drawings are not infrequently ruined by the spilling of ink, which might have been averted by the use

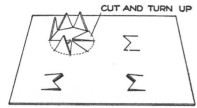

CUT AND TURN UP

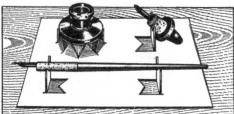

A Sheet of Heavy Paper Quickly Transformed into a Holder for Ink Bottle and Pen

of the simple inkstand cut from a sheet of heavy paper, as shown in the

sketch. The upper illustration shows the method of cutting the paper to fit the ink bottle and stopper, and to produce a pen rack. The device will be found especially useful when materials for drawing are used away from a place especially fitted for the purpose, since the inkstand can be made in a few minutes from material readily available.—Contributed by Henry C. Franke, Jr., Philadelphia, Pa.

How to Wind Wire on Electrical Apparatus

When a beginner, it was the despair of the writer to try to produce in his homemade apparatus the mathematical regularity and perfection of the winding on the coils of electrical instruments in the supply stores, but when he found that this professional and workmanlike finish could be obtained by means of a simple contrivance, and a little care and attention to details before beginning, experimental work took on a new interest.

At the outset let it be stated that wire should never be wound directly on the iron core, not only because it cannot be done satisfactorily in that manner, but for the reason that it is often desired to remove a coil from a piece of apparatus after it has served its purpose. It is therefore advisable to make a bobbin, which consists of a thin, hard tube with two ends. The tube may be easily formed by wrapping a suitable length of medium-weight paper on the core, having first coated it with ordinary fish glue, excepting, of course, the first 2 or 3 in. in direct contact with the core. Wind tightly until the thickness is from $\frac{1}{32}$ in. to $\frac{1}{16}$ in., depending upon the diameter of the core, and then wrap with string until the glue hardens, after which the tube may be sandpapered and trimmed up as desired.

Where the wire is not of too small a gauge and is not to be wound to too great a depth, no ends will be necessary if each layer of wire is stopped one-half turn before the preceding one, as indicated in the accompanying sketch, and is also thoroughly shellacked. With ordinary care magnet wire may be wound in this manner to a depth of over one-half inch.

The tube having been made ready, with or without ends as may be necessary, the small winding jig illustrated is to be made. All that is essential is to provide a suitable means for rotating by hand a slightly tapering wood spindle, upon which the tube is to be pushed. The bearings can be just notches made in the upper ends of two standards, through each of which a hole is drilled at right angles to the length of the spindle, so that some string or wire may be laced through in order to hold the spindle down. A crank may be formed by winding a piece of heavy wire around the larger end of the spindle. A loop of wire, or string, is to be attached at some convenient point, so that the crank may be held from unwinding while adjusting matters at the end of each layer, or while making a connection. There should also be provided a suitable support for the spool of wire, which is generally placed below the table to good advantage. Much depends, in this sort of work, upon attention to these small details, after which it will be found that

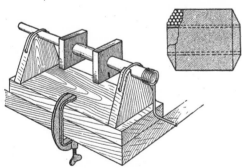

Winding a Coil of Wire so That the Layers will be Even and Smooth

the actual winding will require very little time.—Contributed by John D. Adams, Phoenix, Ariz.

⸿A No. 10 gauge shotgun cartridge shell telescoped with a No. 12 gauge shell forms a convenient match safe for campers, or other persons out of doors, and is moisture-proof.

Hourglass Sewing Basket

Two oblong peach baskets, their bottoms fastened together and the whole covered with silk, formed the

Artistic Effects may be Produced by the Inventive Woman

hourglass sewing or darning basket shown in the sketch. Square plum baskets and other forms trimmed in cretonne, linen, or inexpensive goods, depending on the intended use, may also be utilized. Ornamental details may be added to suit the individual taste.

The basket was made as follows: The peach baskets were wired together at their bottoms. A piece of silk was cut, wide enough to reach from the top to the bottom of the joined baskets and to permit the folding over of a portion at the top and bottom. One long edge of the piece was glued to the inner edge of the bottom and drawn in around the sides to form neat folds. The upper edge of the silk was then glued in the top, being folded over the edge.

A cord was fixed around the middle of the basket, as shown in the sketch.

A lining was glued into the top and bottom. It was folded and stitched along its edges to prevent raveling and to give a smooth finish. The bottom need not be lined, but it is desirable to have it so.

The pincushion was made by padding a block with cotton and then covering it with silk. A cardboard box may be used instead. The cushion was nailed into place from the bottom. Ribbon may be used to draw the silk to the sides of the basket at the middle, and a cushion may be made entirely of cotton or cloth and attached with ribbons. —Contributed by Thomas J. Macgowan, Mount Vernon, N. Y.

A Perpetual-Motion Puzzle

The fallacy of perpetual motion is now so generally understood that the description of a new scheme for attain-

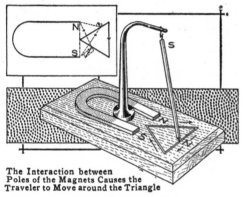

The Interaction between Poles of the Magnets Causes the Traveler to Move around the Triangle

ing it is only justified in so far as it may be instructive. The sketch illustrates such a device, apparently successful, and the discovery of the error in it is both instructive and interesting.

Mount a horseshoe magnet on a wooden base, and into the latter cut a continuous groove along the three sides of a triangle opposite the poles of the magnet, N and S. Suspend a long, narrow bar magnet on a universal joint from a standard. A pin projects into the groove from the lower end, which is its north pole, and can move only along the triangular course.

Start the device with the suspended magnet in the position shown. The lower end will tend to move in the direction of the arrows, because in so doing it is getting farther away from the repelling north pole of the horseshoe magnet and nearer the attracting south pole, which action will bring it to the corner of the triangle in the foreground. It will next move down the side as indicated by the arrow, because along that line it is nearer the attracting south than the repelling north pole. When it reaches the end of its trip, at the angle between the poles of the magnet, the attraction and repulsion will be balanced, but a slight jar will carry the traveler beyond the angle.

The third leg of the triangle will be covered similarly, the north pole repelling the traveler. On this basis the motion should continue indefinitely, but a test will show that it will not do so.

The corners of the triangle should be rounded slightly and it would be better to use several hanging magnets, flexibly connected, so that when one is at the dead center the others will carry the traveler on.

How to Transfer Drawings

Soiling of drawings transferred with carbon paper may be avoided by substituting a piece of unfinished paper, the surface of which has been covered with a thin coating of lead rubbed from the pencil. If any errors are made in the tracing, or undue pressure is applied with the hand, the resulting impressions may be removed readily with an eraser.

If a copy of a drawing is desired, and it is not necessary that the same relative left and right position be maintained, the original pencil drawing may be placed face downward on a sheet of paper and the back of it rubbed with a bone paper knife, or other smooth, rounded object. By going over the impression and making a reverse of it in the same way a copy of the original in the same relations may be obtained.—Contributed by J. E. Pouliot, Ottawa, Canada.

Pivoted Searchlight Made of an Old Milk Strainer

By JOHN J. SPAULDING

BOTH as a safety device and a practical novelty, a homemade searchlight for a canoe, or other small craft, is worth while making. An old milk strainer was used for the reflector of that shown, and many equally serviceable makeshifts can be devised easily from old cans, or formed from sheet metal. The detailed construction, as indicated, is suggestive only, since local conditions and materials available will govern the design of the fittings. The light has a double control, one cord governing the vertical adjustment, and another, arranged like that of a ship's wheel, the horizontal pivotal range. The vertical adjustment is desirable, but not essential. The control cords are run in screw eyes along the coaming of the craft, so that

Canoeing and Boating at Night Is Safer and More Pleasurable if the Craft is Equipped with a Searchlight. This One was Made of Pick-Up Materials at Small Cost

MILK STRAINER
WOOD BLOCK
PORCELAIN SOCKET
SCREW EYE
VERTICAL-CONTROL CORD
LIGHT CORD
CONTROL CORD

The Support for the Reflector is Pivoted in the Deck, Reinforced as Indicated

cells, stored under the bow deck, or in a box set at some other convenient place, supply the current for the 6-volt lamp.

The main dimensions of the fittings, as detailed, are: strainer, 10 in. in diameter and 10 in. long; vertical support, 12½ in. over all, and 6 in. wide at the upper portion; the wood used is ½ and ¾ in. thick, except that for the pivot post, which is 1 in. thick. The reflector is fitted with a wooden block through which the porcelain socket is set, as shown. A knife switch,

one person can paddle the canoe, and adjust the searchlight as well. Dry

placed near the stern of the craft, controls the connection with the battery circuit.

The inside of the reflector should be polished with emery cloth, and if the surface is rough, it may be painted with white enamel. The outer surfaces of the metal part are painted black. The wooden parts may be painted, or given several coats of spar varnish, to withstand the weather.

Gravity-Feed Coal Hopper on Truck

In the large farm kitchen, in the workshop, and even for firing a small furnace, a coal hopper that will hold considerable coal, and that can be rolled along the floor easily, is a convenience. Such an arrangement, made from a section of galvanized-iron pipe, 10 in. in diameter and 30 in. long,

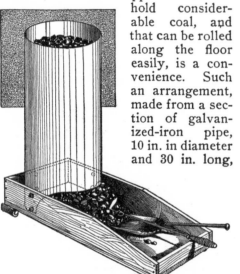

The Large Capacity of the Hopper and the Ready Portability of This Arrangement Are Practical Features

is shown in the sketch. The pipe was cut at one end, as shown, so that when the coal was poured into the hopper, it would feed out. A truck, mounted on casters, was made, 12 in. wide, 5 in. high, and 25 in. long. The hopper was mounted on one end of it, and bolted securely at the sides and end. The coal is shoveled into the hopper at the bin, and the load pushed to the furnace, where it can be easily used as needed. The construction can be made larger for use with a coal scoop, in firing a boiler or large furnace.—L. R. Markwood, Factoryville, Pa.

Taking Photographs in Falling Snow

Falling snowflakes in a camera photograph—the large feathery, slow-falling kind—often make an exquisitely beautiful picture out of a commonplace scene. And while the great majority of the attempts to get them prove failures, the photographer—usually an amateur—needs only to provide an avenue in front of his lens a short distance, that is free from falling flakes, by the use of a shelter such as a tree or porch. The slow snapshot necessary in cloudy weather will not stop the motion of the flakes nearest the camera, and these passing through greater angles of space in equal length of time than those farther away, will blur on the negative. We made some excellent outdoor views in deep snow, while heavy snow was falling, with flakes splendidly decorating the darker regions of figures and foliage, by holding a felt hat and two umbrellas in a line in front of the camera, and above the range of the lens. — J. Cecil Alter, Cheyenne, Wyo.

A Double-Contact Vibrator

A double-contact vibrator, which eliminates sticking contacts, spring troubles, and other sources of annoyance, in addition to producing a fine, high tone, is shown in the sketch. It is an instrument easy to construct, by reason of its simplicity. Special care in making the vibrator D will insure good vibration. The springs, holding the contacts, are of phosphor bronze. The contacts may be made of silver, platinum, or other metals, which will not burn and break contact. The coils

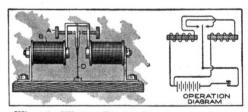

When the Vibrator Touches One Contact, the Coil on the Opposite Side Attracts the Vibrator, This Process being Repeated Alternately

B are of the common bell-ringing type. The springs on the vibrator should not be too long, nor too weak; experimenting will determine the length at which they will work best. The adjustment is made at the thumbscrews A. The coils are supported on metal brackets, bolted to a wooden base. The method of hooking up the vibrator in the key circuit is shown in the diagram.—J. L. Taylor, Barker, N. Y.

Battery Buzzer Converted into a Telegraph Sounder

An ordinary battery buzzer may readily be converted into a telegraph sounder for use in practicing the Morse code. All that is necessary is to connect the vibrator contact C of the buzzer to the binding post that is not insulated from the frame. The other connections of the key and battery are the same as in any ordinary telegraph or buzzer circuit. In the diagram, C represents the vibrator contact; D, the wire connecting the contact and the

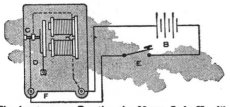

The Amateur can Practice the Morse Code Handily on This Sounder, Made from a Buzzer

uninsulated binding post, and F, the uninsulated binding post; E is the telegraph key, and B, the dry cells.—Clarence F. Kramer, Lebanon, Ind.

Lawn Seats Built on Tree Stumps

A practical use to which stumps, left from the felling of trees, are put in a city park is as supports for lawn benches. This obviates the need of grubbing them out, while the work of preparing them to receive the seats is less than would be required to remove the stumps. Of course, the location of the stump will determine whether it is worth while as a support for a seat, or had better be grubbed out. Many designs are possible, and the position and size of the stumps available will suggest suitable construction. The

These Practical Lawn Seats Show the Possibilities of Stumps as Supports for This Purpose. The Conversation-Chair Design Is Especially Interesting

woodwork for the seats is simple, and the benches can be made removable for the winter if desired.—C. L. Meller, Fargo, N. D.

A Hinged Box Cover Made without Hinges

When a wooden box with a hinged cover is desired, especially a small one, hinges are not always at hand, and are sometimes difficult to obtain. Under these circumstances a good substitute is to make the box as shown in the sketch, using the extension of one end and nails, or screws, driven through the back corners of the lid, as a hinge arrangement. This makes the use of hinges unnecessary, and is serviceable even for permanent use. Where hinges projecting from the surface of the box are objectionable, this method of construction is especially desirable. It is best to make the hinged ends with the grain vertical, and to round off the hinged corners of the lid slightly.—R. J. Rohn, Chicago, Ill.

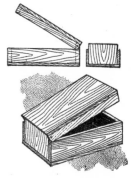

Automatic Flash Light Snaps Chicken-Coop Marauder

After the wire fence around the chicken house had been torn up, and

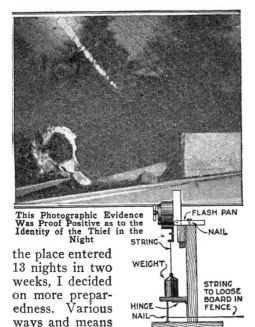

This Photographic Evidence Was Proof Positive as to the Identity of the Thief in the Night

the place entered 13 nights in two weeks, I decided on more preparedness. Various ways and means failed, so I used a comparatively slight knowledge of photography in the process.

I mounted my flash lamp on a piece of board, 1 by 4 by 8 in. long, and fastened this to a base, as shown. I attached a weight to the lamp, which was supported by a hinged drop, halfway down the upright board, which in turn was supported by a nail, to which was attached a string. The flash was set off by a slight pull of the string, which dropped the weight. This contrivance I concealed in the chicken yard, and the camera in the chicken house. That night I opened the lens of the camera in the dark, and attached the string to a loose board in the fence. The next morning, before daybreak, I closed the lens again. The flash had been set off during the night. Also there were drops of blood on the ground. I could hardly wait until the plate was developed. The result, as reproduced, was hardly what I expected.—H. U. Scholz, Medford, Ore.

A Fishing-Tackle Outfit in a Shotgun Shell

At the camp or on the trail, an emergency fishing-tackle outfit is almost as handy as matches, compass, and knife, and it may even be the means of saving one's life. A convenient way to carry such an outfit is in two old shotgun shells, telescoped. The hooks, on a cork, and the sinkers are fitted snugly into the shell. Several yards of line are then wound on the outside. This outfit can be stowed into a pocket handily, always ready for use.—E. Everett Buchanan, Jr., Elmira, N. Y.

A Split-Bamboo Lettering Pen

Marking of packages and similar lettering can be done neatly with a pen made in a few minutes from split bamboo fitted with a short section of watch spring. Select a piece of bamboo, 1/4 by 1/16 in. and about 7 in. long, and finish the end, as at A. Trim the end to an angle, as at B, and then point it, as at C. Split the point carefully, as at D, and smooth away the tufts at the edges. Cut a piece of watch spring the width of the pen point and bind it into place, arched as shown. To use the pen, insert ink into the arch of the

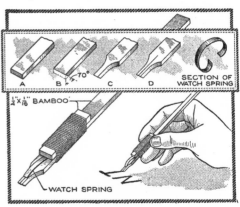

This Pen, Cut from a Piece of Bamboo and Fitted with a Spring Fountain Device, Is Especially Useful for Marking Packages

spring, and it will work much like a fountain pen.—Raymond H. Lufkin, Dorchester, Mass.

How to Make a Houseboat

By H. SIBLEY

THE houseboat shown is of the scow design, 6 ft. wide by 20 ft. long, with the cabin extending beyond the scow 1 ft. on each side. The scow tapers up at the forward end and is protected with a heavy sheet-iron plate so that the craft may be snubbed up on sandbars without danger of springing a leak, even though a submerged log be struck while running at full speed.

The power plant consists of a standard 4-hp. reversing gasoline engine which drives the paddles at their most efficient speed, 45 revolutions per m i n u t e through a 13-to-1 reduction. Cast-iron hubs, into which are inserted cold-rolled steel spokes, and wood paddles bolted to their ends constitute the propeller wheels. The cruising speed is about 4 miles an hour.

Two wide bunks, beneath which is locker space, provide sleeping accommodations for a crew of four. In the kitchen the motor and gearing are almost completely concealed under the work

table. The cooking is done on a two-burner blue-flame kerosene stove, and the sink is provided with running water

The Hull of the Houseboat is Built on the Scow Type so That It can be Run in Shallow Water without Danger

suitable for washing dishes, etc. This water is drawn from a 30-gal. tank on the roof, which is filled by a centrifugal pump driven from the engine shaft. A modern toilet room is installed, and an ice chest on the after deck will hold supplies and ice for a week's cruise.

An acetylene-gas lighting system is installed and is used to light both cabins and a searchlight. A heavy anchor of special design is manipulated by a windlass on the forward deck. A

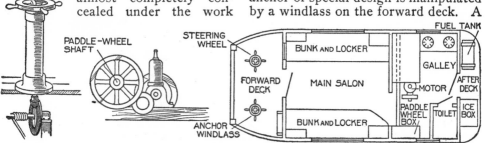

Detail of the Anchor Windlass and Engine Gearing; Also the Deck Plan, Showing the Location of the Parts and the Arrangement of the Cabins

similar device controls the rudder. Life rafts, complete with paddles, are placed on the roof, and in hot weather these are moved to one end and an awning erected to make a cool sleeping place.

Wood Box with a Refuse-Catching Drawer

The ordinary wood box may be greatly improved by adding a drawer at the top and one at the bottom, as

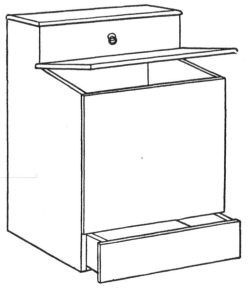

The Wood Receptacle Has a Drawer Bottom for Catching the Dirt, Which can be Easily Cleaned

shown in the sketch. The upper drawer is used for storing the flat-irons and stove-polishing materials, and the lower drawer is the bottom of the wood box. The dirt and pieces falling from the wood remain in the drawer, which can be removed and cleaned easily.—Contributed by William Jutila, Astoria, Ore.

Disappearing-Coin Trick

To make a quarter disappear from a glass of water after hearing it drop is a very puzzling trick. The articles necessary to perform this trick are a glass of water, a handkerchief, a quarter and a piece of clear glass the exact size of a quarter. The glass can be cut and ground round on an emery wheel, and the edge polished.

To perform the trick advance with the piece of glass hidden between the second and third fingers of the left hand and holding the quarter in plain sight between the thumb and first finger of the same hand and the handkerchief in the right hand. Throw the handkerchief over the left hand and gather up the glass piece in the fold of the cloth, allowing the coin to drop into the palm of the left hand while covered. Remove the left hand and hold out the piece of glass with the handkerchief drawn tightly around it. Anyone can touch the cloth-covered glass, but it cannot be distinguished from the quarter. While this is being shown slip the quarter into a pocket. Spread the handkerchief over the glass of water and allow the glass disk to drop. A distinct click will be heard when it strikes the bottom. Raise the handkerchief and nothing will be seen, as the glass will not be visible in the water.—Contributed by Amon H. Carr, Gainesville, Tex.

Watering Window-Box Flowers

The window box for flowers can be conveniently watered in the following manner: Construct a metal box to receive the box holding the soil and bore enough holes in its bottom to admit water to the soil. The inside box should be supported about 2 in. above the bottom of the metal box. Sponges are placed in the bottom to coincide with the holes in the soil box. A fill-

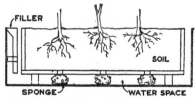

The Soil is Kept Moist by the Water Feeding through the Sponges from the Under Side

ing tube is made at the end. The water is poured into the metal box and the sponges admit only enough water for the plants at all times.

How to Make Combined Kites

BY C. M. MILLER

PART I—A Dragon Kite

DRAGON kites are made as hideous as the maker can possibly conceive, and while the one to be considered is no beauty, it is more droll than fierce-looking. In general appearance the dragon and centipede kites are like huge caterpillars floating about in the air. The kite sometimes twists and the balancer sticks appear to be large hairy spines. Usually the tail end swings higher than the head. It is like so many single kites, pulls hard, and requires a strong cord for the line. The individual circular sections may number 20, and if placed 30 in. apart, would make a kite about 50 ft. in length, o r t h e number of sections may be more or less to make it longer or shorter. The kite will fold

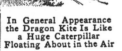

In General Appearance the Dragon Kite Is Like a Huge Caterpillar Floating About in the Air

up into a very small space, for carrying about or for storage, but care should be taken in folding not to entangle the harness.

The Head

The head requires much more work than any of the other sections. There are two principal rings to this sec-tion, as shown in Fig. 1. The inner ring is the more important, the outer one being added for the protection of the points when alighting. The construction of the framework is shown in Fig. 2. It is made entirely of bamboo. The bamboo is split into strips, about $\frac{3}{16}$ in. wide, for the ring A. As the bamboo strips will be much too thick, they must be pared down to less than $\frac{1}{16}$ in. The diameter of the ring A is 12 in., and a strip of bamboo to make this ring should be about 38 in. long, so that there will be some end for making a lap joint. The ends of the strip are held securely together by winding them with linen thread. The Chinese boys use strips of rice paper, about $\frac{1}{2}$ in. wide and torn lengthwise of the paper. The rice-paper strips are made wet with paste before winding them on the joint, and when they dry out the shrinkage will bind the ends securely.

Two crosspieces, of the same weight as the ring stick, are placed $3\frac{1}{2}$ in. apart, at equal distances from the center and parallel, as shown at B and C. The ends of these pieces are turned at a sharp angle and lashed to the inside surface of the ring A. To make these bends, heat the bamboo over a candle flame until it will give under pressure, then bend and it will stay in shape after it becomes cold. This method of bending should be remembered, as it is useful in making all kinds of kites. Two small rings, each 3½ in. in diameter, are put in between the two parallel pieces, as shown at D and E. These are for the eyes of the dragon. The

FIG. 1

The Kite-Head Section, Having Horns, Ears, and Revolving Eyes, Is Very Hideous

rings are lashed to the two crosspieces B and C. As the eyes revolve in the rings they should be made perfectly true. This can be done by shaping the bamboo about a perfectly round cylinder, 3½ in. in diameter. To stiffen the whole framework, two pieces of bamboo, 1/16 in. thick, ⅛ in. wide, and 20 in. long, are lashed to the back as shown by F and G. There is a space of 3 in. between the inner ring A and the outer ring H, giving the latter a diameter of 18 in. It is made of a bamboo strip, ⅛ in. wide, and should be less than 1/16 in. thick. It may be necessary to make this large ring from two pieces of bamboo, to get the length, and in such case be careful to make a perfect ring with the ends well lashed together. Two short pieces are lashed to the two rings, as shown at J and K. The supports for the horns consist of two pieces, ⅛ in. wide and less than 1/16 in. thick, and they are lashed to the upper crosspiece and to both rings, so that the parts L and M are exactly halfway between the ends of the pieces F and G and radiate out from the center of the ring A, the other parts, N and O,

pointing to the center of the eye rings, respectively. The ears are unimportant and may be put on if desired. The rings on the horns and the stick ends may be from ½ to 2 in. in diameter, cut from stiff paper, but if larger, made of bamboo.

Chinese rice paper is the best material for covering, and it should be stretched tightly so that there will be no buckling or bagging places. The only part covered is that inside of the inner ring A, the horns, and the ears, leaving the eye rings open. The shades are put on with a brush and water colors, leaving the face white, or it can be tinted in brilliant colors. Leave the horns white and color the tongue red.

The Eyes

The frame for each eye is made of bamboo, pared down to 1/32 in. in thickness and formed into a perfect ring, 3¼ in. in diameter. Each ring revolves on an axle made of wire passed through the bamboo exactly on the diameter, as shown at P, Fig. 3. The wire should be long enough to pass through the socket ring D or E, Fig. 2, also, and after the eye ring is in place in the

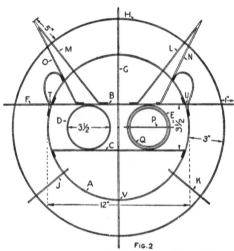

FIG. 2

The Framework for the Head Section is Made Entirely of Bamboo Strips Lashed at the Joints

socket ring and the axle adjusted, the latter is fastened to the eye ring with a strip of paper wrapped tightly around the wire and pasted to the bamboo of the ring. A glass bead, placed on the wire axle between the socket rings D or E and the eye ring Q on each side, keeps them apart and the revolving one from striking the other.

Each side of the eye ring is covered halfway with rice paper, as shown in Fig. 4. The part R is on the upper front half, and that shown by S is on the back lower half. Placing the two halves in this manner causes an unequal pressure of the wind on the whole eyepiece, and thus causes it to revolve on the axle. The front upper half of the eyepiece is made black, and the smaller dark portion extending below the darkened half is a round piece of paper placed just between the two halves so that half of it will show on both front and back of the eyepiece. When the eyepiece is given a half turn in its socket the back side will come to the front and will appear just the same as the other side. Some kite builders add pieces of mirror glass to the eyes, to reflect the light and cause flashes as the eyes revolve in their sockets.

A Section Kite

The ring for the section kite is made the same size as the inner ring of the head kite, or in this case 12 in. in diameter. The bamboo for making this ring should be ⅛ in. wide and 1/16 in. thick.

be made small, light and well balanced. Small tufts of tissue paper, or feathers, are attached to the tip ends of the balancer sticks, as shown in Fig. 5. The

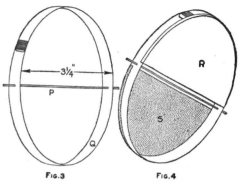

Two Bamboo Rings with Paper Coverings, to Make Them Turn in the Wind, Constitute the Eyes

cover for the section kite is put on tightly, the same as for the head, the colors indicated in the sketch being only suggestions, as the builder can color them as desired. The balancer on the last section should have streamers, as shown in Fig. 6, for a finish. The streamers are made of light cloth.

The Harness

As previously stated, 20, more or less, sections can be used, and the number means so many separate kites which are joined together with three long

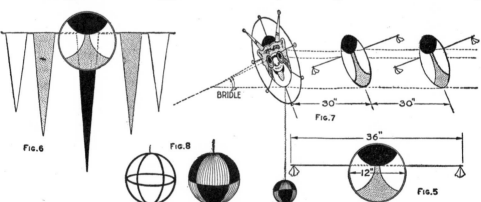

The Section Kites Have Balancers and the Disks are Given Bright Colors So That They will Produce the Effect of a Great Dragon When in the Air, the Head Piece Having a Ball Balancer Hanging from the Under Edge

The balancer stick, 36 in. long, is located about the same place as the cross-stick F, as shown in Fig. 2, and must

cords, spacing the sections 30 in. apart. The cords should be as long as the kite from the head to the tail, allowing sufficient extra length for the knots. As such a kite will make a hard pull, the cord used should be a six-ply, hard-twisted seine twine. Start by tying the three long cords to the head kite at the points T, U, and V, Fig. 2. Tie the next section at corresponding places just 30 in. from the head kite. The construction will be much easier if the head kite is fastened to a wall so that each cord may be drawn out to its proper length. Continue the tying until all sections are attached just 30 in. apart. Other spacing can be used, but the distance selected must be uniform throughout the length of the kite. The individual kites, or sections, may vary in size, or they can all be 9 in. in diameter, instead of 12 in., and the balancer sticks 30 in. long instead of 36 in., but a kite of uniform sections is much better and is easier to make. The positions of the sections as they will appear in the kite are shown in Fig. 7.

The Bridle

The Chinese bridle is usually made of three strings, which are attached to the same points on the head kite as the harness cords, or at T, U, and V. The lower string is longer than the two upper ones so that the proper inclination will be presented to the breeze. As the head is inclined, all the section kites will also be inclined. Some makers prefer a balancer on the head kite, and in one instance such a balancer was made in the shape of a ball. A ball made of bamboo strips is shown in Fig. 8, and is attached as shown in Fig. 7.

Flying the Kite

It will be necessary to have a helper, and perhaps two, in starting the kite up, as the harness might become entangled. Quite a little run will be necessary, but when up the kite will make a steady flier and will pull very hard. If the first attempt is unsuccessful, try readjustment of the bridle or a little different position in the breeze, and see that the balancers are not tangled. Quite a number of changes may be worked out on these plans, but it is necessary to bear in mind that the distances between sections must be equal and that the general construction must be maintained.

A Movable Sunshade and Seat for Garden Workers

Weeding gardens under a hot sun and in a necessarily uncomfortable position is by no means a pleasant occupation, but much of the hardship can

The Sunshade and Seat are Mounted on Wheels So That the Device can be Easily Moved About

be avoided if a combination movable sunshade and seat is made, as shown in the illustration. The framework consists of two end pieces, both made in the same way. Two ½-in. tees are required for each. These are connected with a 5-ft. pipe, for the upright, and the crosspiece that supports one end of the canvas consists of two pieces of pipe, each about 1½ ft. long, screwed into the tee. The axle for the wheels is also formed of two pieces of pipe, but these are only about 1 ft. long.

Four suitable cart wheels should be provided, or solid circular disks cut out of a board or plank, and drilled for a ⅜-in. axle. Ordinary ⅜-in. washers are placed on each side of the wheels, one serving as a shoulder on the pipe end, and the other for the nut. The seat can be made of a 1 by 8-in. board, about 6 ft. 9 in. long, notched at each end to fit the pipe tees and prevented from shifting by means of U-bolts around the axle on either side of the upright. After covering the top with canvas, fastened at each end around the upper crosspiece, the shaded movable seat is ready for service.—Contributed by W. E. Crane, Cleveland, Ohio.

To Keep Grass and Weeds Out of Tennis Courts

Tennis courts are very apt to become covered with grass and weeds unless considerable labor is expended on them. The best way to keep a court free from this growth is to put on sheets of tar paper close together when it is not in use. The paper should be cut in strips the proper length, so that they can be rolled up and set aside until the game is over, when they can be laid down again. After the court has been covered a few times, the grass will die out and disappear altogether. The use of salt, acids, and a number of other things, together with hoeing, were tried out, but the paper covering was the most successful of them all.—Contributed by W. D. Mills, Bakersfield, Cal.

Buttonhole-Bouquet Holder to Keep Cut Flowers Fresh

Being employed near a glass-blowing department I procured a small glass  tube and made a buttonhole - bouquet holder. One end of the tube I closed and flared the other, then flattened the whole tube a bit. This tube, inserted in the buttonhole with a few drops of water in it, will keep cut flowers fresh for a whole day. If the upper end is flared just right it will fit the buttonhole and will not be seen, as the tube is hidden beneath the coat lapel.—Contributed by Frank Reid.

Cooler for a Developing Tray

Regardless of the fact that tank development for photography is the better method under most conditions, there

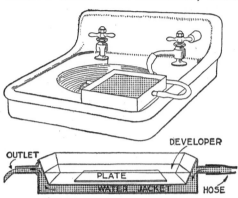

Water-Jacketed Tray for Use in Cooling a Developing Tray with a Flow of Water

are many who take special delight in tray development, because they enjoy seeing the picture as it develops on the plate or film.

There are many of this class of photographers both among amateurs and professionals, and they universally experience much trouble in keeping the solutions cool enough for good work in hot weather or a warm dark room. Many schemes have been used, but there is none simpler than that shown in the illustration. It is certainly better than letting the developing tray float in the bathtub, a common practice among amateurs who are compelled to use the bathroom as a dark room.

Any tinsmith can make a water-jacketed tray of zinc at a nominal price. If one expects to use it for developing films by hand, it should be made relatively deep, or at least shaped so that whatever method is used in manipulating the film will be suited to the size and shape of the tray. I had one made of zinc to accommodate an 8 by 10-in. plate and providing about 1/4-in. space all around the sides and bottom of the inner tray for the circulation of the water.

Two brass gas tips were soldered into the ends of the tray and a rubber tube was attached to one of them, as shown.

By connecting this tray to the bathroom hydrant the developing solution can be kept as cool as desired without slopping water where it is not wanted.—Contributed by T. B. Lambert, Chicago, Ill.

Doorbell Push Button on Screen Door

Push buttons are frequently put in hallways, or other places, not accessible when the screen doors are hooked

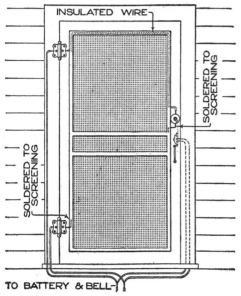

INSULATED WIRE

SOLDERED TO SCREENING

SOLDERED TO SCREENING

TO BATTERY & BELL

Method of Connecting a Push Button for a Doorbell on a Screen Door

shut. By connecting an extra button on the screen door with the regular doorbell line, the service can be made to go on uninterrupted whether the screen be locked or open.

To do this, one wire, carefully insulated, from the outside button is laid under the upper molding strip of the screen, as shown, coming out through the hinge and passing down to the battery line, being concealed in any suitable manner, by molding or within the door frame. The button circuit is completed by connecting it to the screen with a short wire, the screen then acting as a conductor. If preferred, the line may be run down under the screen-door molding, and out through the bottom hinge same as with

the other wire at the top. **If the screen** is used as conductor, a piece of wire should be connected to it near the bottom, and led through the hinge as before. In either case, after leaving the hinge, the wire must be led and connected to the battery line, to complete the circuit. It is then possible to operate the bell either from outside of the screen door or at the regular place within.—Contributed by E. M. Davis, Philadelphia, Pa.

Working Wood by the Application of Heat

It is often desirable to fit a piece of wood into a piece of metal by means of cutting a thread in the metal and screwing the wood therein. This can be accomplished by heating the metal to a little over the boiling point of water and screwing the wood piece into the metal while hot; or, if this is impractical on account of size, to heat the metal, make a screw plate by cutting a thread in a small piece of metal, the size and thread corresponding to the hole into which the wood piece is to be inserted, and heating it to the required temperature, then running the wood with some pressure through the thread. The wood then can be screwed into the larger piece, where it will hold firmly.

The process of heating wood without the aid of steam can be used to advantage in a number of ways; for instance, a hammer handle that is crooked can be straightened by careful heating without burning; also billiard cues, or almost anything of hard wood. It is surprising how easily it is done and how permanent the repair will be. The Indians at one time made their arrows from small hardwood twigs which were almost always crooked to start with, but after being dried they were warmed over a fire and straightened.

Another use for the application of heat is as follows: When it is desired to place a screw in a fragile piece of wood that is likely to split, if the screw is heated to a blue color and turned into the wood while hot, there will be scarcely any danger of splitting. In this case do not try to use oil or a lubricant of any kind, as the screw is sure to set before it is in place.—Contributed by James H. Beebee, Rochester, N. Y.

A Parlor Table

The material required for the parlor table illustrated is as follows:

1 table top, 1 by 26 by 41 in.
1 bottom shelf, 1 by 15 by 35 in.
2 side rails, ¾ by 4 by 33 in.
2 end rails, ¾ by 4 by 21 in.
2 top cross braces, 1 by 4 by 19½ in.
4 feet, 1¾ by 4 by 4 in.
2 posts, 6 by 6 by 26 in.
2 side corner strips, 1 by 1 by 31½ in.
2 end corner strips, 1 by 1 by 17½ in.

The bottom shelf can be made of two pieces of 1-in. material, 8 in. wide, carefully glued together, and reinforced on the under side with two crosspieces, glued and screwed to it. The foot pieces are secured to the bottom shelf so as to project 1 in. on the ends and sides. In case a center support is deemed advisable, another foot piece can be added, but unless the floor is very level, rocking may result. The uprights, or posts, are made from solid 6 by 6-in. lumber, 26 in. long, carefully squared at the ends, and tapered to 4 in. square at the upper end. If desired, the posts can be made of

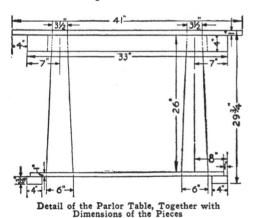

Detail of the Parlor Table, Together with Dimensions of the Pieces

boards, cut and fastened together to form a hollow tapered post. In either case, they should be set in about 4 in.

from each side of the bottom shelf and fastened to it by means of screws.

The rail pieces for the table top

Design of a Table That will Appear Well in the Different Oak Finishes as Well as in Mahogany

should be cut and fitted with mitered joints at the corner to form a rectangular frame, 21 by 33 in. This is glued to the top and may be toenailed to it; but to provide a more secure bracing, a 1-in. square strip of material is fastened all around the inside edge of the rails, flush with their upper edge. The top is screwed to this. In order to prevent tipping when the top is resting on the 4 by 4-in. ends of the posts, two cross braces are provided. These should be screwed to the outer-end sides of the posts, and beveled off on their upper edges to fit the table top. They should be of such length as to have a tight fit between the side rails, and are fastened to these by means of finishing nails driven from the outside. Gluing and toenailing can also be used to secure the top more firmly to the braces; care should be taken that no nails cut through the table top. After thoroughly sandpapering and smoothing off the table, it can be finished to suit.—Contributed by H. J. Blacklidge, San Rafael, Cal.

Homemade Fuses for Battery Circuits

To carry out the general arrangement of the regular electrical equipment of a large power plant, the battery circuit should have some sort of a fuse block. An excellent way to make such a block is as follows: Procure a piece of glass tubing, about 1 in. long, and make a mounting for it with four pieces of sheet brass, as

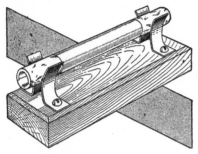

The Fuse is Well Insulated and Protected against Fire the Same as the Large Fuse

shown in the sketch. The brass pieces are shaped and fastened to a wood base so that their upper ends form a clamp to hold the glass tube.

A very thin piece of tinfoil, about 1½ in. long, is cut ⅛ in., or less, in width, the size depending on the amount of current used. This is placed in the glass tube and the ends are bent over the edge. A piece of tinfoil is then wrapped around each end, over the ends of the inclosed piece, and pasted to the tube.

The fuse thus made is pushed into the spring clamps of the block which are connected in the circuit.—Contributed by Charles W. Meinberg, Somerville, Mass.

Reproducing Flowers and Leaves in Colors

A very cheap, easy, and quick way to make reproductions of flowers, leaves, insects, butterflies, etc., is the following: Purchase 1 oz. of bichromate of potash and dissolve it in water. Spread this fluid over the surface of the paper upon which the picture is to be made, using a soft brush, and place it in a dark room to dry. When dry it will be a light lemon color.

The leaf, or part to be reproduced, is placed upon the yellow side of the paper and the whole exposed to the sunlight. The uncovered portions will gradually turn brown, while the part

covered by the object will remain a lemon color. When the desired shade of brown is attained, remove the paper from the sunlight, and the image of the object will be on the paper.

Go over the outline with crayons or colors, and fill in the print according to the natural colors. Very beautiful pictures may be made in this way, and a butterfly made up in natural colors on a dark-brown ground is very pretty. These prints, uncolored, will keep for some time, if they are not exposed to a strong light.

Washing as ordinary photographic prints are treated will improve them a little. If the instructions are followed, many beautiful objects can be preserved in color.—Contributed by J. B. Murphy, Plainfield, N. J.

Dishwasher and Drier

Where hot water is available the dishwasher illustrated is especially suitable. It is easily constructed and inexpensive, the only real expense being for the tank, which is 15 in. deep and 20 in. in diameter. A washer of these dimensions will hold the dishes from a table serving six to eight per-

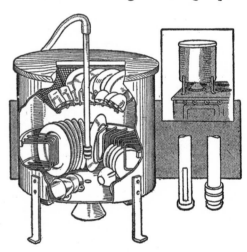

The Hot Water from the Faucet is Forced over the Dishes in a Fine Spray

sons. The tank may be supported on legs if desired.

The supports for holding the dishes

in a vertical position are made of wire in a U-shape, the ends being fastened into two wood hoops that snugly fit the inside of the tank, one near the bottom and the other about 9 in. above the first one. The wires are set about 1½ in. apart and radiate from the center like wire spokes in a wheel.

A funnel-shaped piece, having a hole 5 in. in diameter in the center, is made of mesh wire and hung in place with several wires running to the upper edge of the tank. Hooks are attached to the edge of the hole on which cups and glasses are hung.

The cover consists of a wood disk, with a hole bored in its center for a ½-in. pipe. The piece of pipe used should be 3 or 4 in. longer than the depth of the tank. A long thread is cut on the lower end of the pipe, and two rectangular holes are cut in the pipe end through the threaded part. These holes are made by cutting through the pipe as if making a slot for a key. Two large nuts are run on the threads of the pipe. These nuts should be thick enough to cover the slot in the pipe. A pipe cap is then screwed on the end of the pipe. The upper end of the pipe is attached to a hose connected to the hot-water faucet. By separating the two nuts just a little, a fine spray of water will be forced out of the holes.

When the dishes are in place the spray of hot water can be directed over them by moving the pipe up and down through the hole in the cover.

After the dishes are washed sufficiently the hose is removed and the tank placed over a gas jet so that the heat will pass up through the funnel-shaped attachment in the bottom of the tank. The dishes, already heated by the hot water, soon dry in the heat from the gas jet. If desired, the tank can be allowed to remain over the jet until the next meal is ready to serve, using a very small flame.—Contributed by W. K. Baxter, Massillon, O.

⁋To empty a large sack filled with heavy material, turn or roll it over a barrel.

How to Make Pop-Corn Cakes

It is very difficult to take a bite from a ball of pop corn, and it becomes more difficult as the ball increases in size.

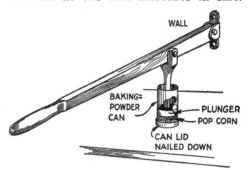

As a large number of balls were required for a church entertainment I decided to make the pop corn into cakes. This was more easily accomplished than first imagined with the use of a cake-forming device as shown in the illustration. The body of the former was made of a baking-powder can with the bottom removed. The cover of the can was nailed to the top of an old table with its flange upward. A plunger of wood was made to fit snugly inside of the can and a lever, about 3 ft. long, attached to it and fulcrumed to the wall.

After the pop corn has been prepared with the sirup, it is placed in the can and compressed. The can is then lifted out of the cover and the pop-corn cake removed. This method offers a much better way to serve pop corn than in balls. In making the cakes, the can, cover, and plunger must be kept well covered with butter.—Contributed by Howard A. Hopkins, Youngstown, New York.

Threading a Darning Needle

Having occasion to use a darning needle, the eye of which was too small to insert the yarn in the usual manner, I tried the following method with good results. A thread was doubled and passed through the eye, and the loop thus formed was used to pull the yarn through.—Contributed by Joe Henderson, Eldred, Ill.

A Fish Scaler

All kinds of devices, both simple and complex, have been made and patented for use in scaling fish, but for a novelty I found the following, which necessity compelled me to improvise on an outing trip, to be as efficient as any of them. As usual, the commissary, in making up the outfit, neglected to take the curry comb to clean the fish, and at the same time remembered to take a plentiful supply of bottled goods. Long before it became necessary to scale any fish enough bottles had been opened to provide the basis of a tool for the purpose, which I constructed by using the small tin bottle caps, a few being nailed on a block of wood, about 3 in. wide by 4 in. long, making a splendid fish scaler, as good and efficient at home as in the camp, and both

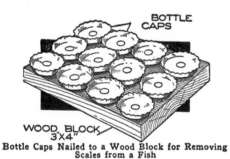

Bottle Caps Nailed to a Wood Block for Removing Scales from a Fish

inexpensive and easily made. The sketch shows the general appearance.—Contributed by T. B. Lambert, Chicago.

A Disappearing Towel

Nothing is more unsightly to a stranger entering a home than a dirty towel in either the bathroom or the kitchenette. To keep the towel out of sight I made a hanger as shown in the illustration. A wire was bent into shape similar to a clothes hanger and a sliding clip made to hold the towel in place. A cabinet was made to accommodate the towel, and the hanger was attached to a cord run over a pulley fastened at the top, through a weight pulley, and then tied to a screw

eye at the top. The weight draws the towel into the cabinet. Near the bottom edge a slot was cut and a small

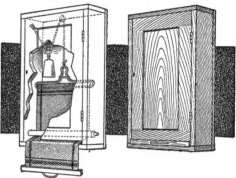

The Weight Draws the Towel into the Case Out of Sight When Not in Use

panel fitted in it. This small panel is fastened to the lower end of the towel. It is only necessary to pull out the small panel to get at the towel. When through with the towel, let loose of it and the weight will draw it into the opening.—Contributed by Chas. C. Bradley, W. Toledo, O.

Ammonia-Carrying Case for Insect Bites

An old clinical-thermometer case can be easily turned into a vial in which to carry ammonia for insect bites. Fit a small rubber stopper in the case, then push a darning needle into the stopper so that its end will be a little more than midway in the case. Cut or break off the needle end projecting on the outside and attach a small wad of cotton to the inside end. The case is then filled with ammonia. For bee stings this works fine, as the ammonia completely neutralizes the formic acid which the bee deposits.—Contributed by E. Everett Buchanan, Elmira, N. Y.

❡The contact points of a firm-joint caliper should never be struck on hard surfaces to adjust them.

How to Make Combined Kites

By C. M. MILLER

PART II—A Festooned Kite

MORE than one kite on the same framework is known as a compound kite. The one illustrated consists of three tailless kites on one long stick, called the spine. The upper one is 3 ft.; the center one, 2 ft., and the lower one, 1 ft. in width. There will be needed for the construction of this kite a stick of light wood—spruce is best, but it may be of pine or bass— 7 ft. long by ¼ by ½ in. If the wood breaks easily it will be better to increase the width from ½ in. to ¾ in., or the stick might be made ⅜ in. thick without increasing the width, but with a good spruce stick the dimensions first given will be sufficient. The stick should be straight-grained and without a twist. If the spine is twisted, the kites will not lie flat or in a plane with each other, and if one is out of true, it will cause the kite to be un-

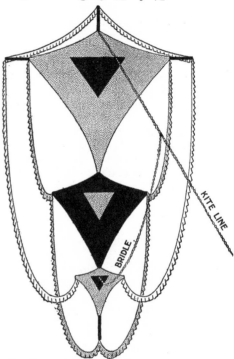

The Kite as It Appears with the Festoons Hung to the Ends of the Sticks

by ¼ by ½ in.; the center one, 2 ft. long by ¼ by ⅜ in., and the lower one, 1 ft. long by ¼ by ¼ in. About five sheets of tissue paper will be required, but more may be needed for color combinations. The so-called French tissue paper is much better, as it comes in fine colors and is much stronger than the ordinary tissue. It costs a trifle more, but it pays in making a beautiful kite. The Chinese rice paper is the strongest, but it comes only in natural colors.

It will be seen that the kites do not extend to the top and bottom of the spine stick. The first bow stick is placed 13 in. from the top end of the spine, and each of its ends extends 6 in. beyond the kite for fastening the festoons. The bow sticks should be lashed to the spine, not nailed. Wind diagonally around the two sticks, both left and right, then wind between the two,

The Spine with the Bow Sticks Properly Spaced as Shown by the Dimensions

steady in the air. The bow sticks are three, the upper one being 4 ft. long

around the other windings. This draws all windings up tightly to prevent slipping.

To string up the upper kite, drill a small hole through the spine, 6 in. from the top, at A, and also 6 in. from each end of the bow stick, at B and C. If a small drill is not available, notch the stick with a knife or saw to hold the string. Another hole is made in the spine 29 in. from the upper bow stick, or at D. Tie the outline string at A, then pass through the hole at C, then through D, up through B and back to the starting point at A. In tying the last point, draw up the string tightly, but not enough to spring the spine or bow. Measure carefully to see if the distance AC is the same as AB, and if CD is equal to BD. If they are not, shift the string until they are equal and wind at all points, as shown at E, to prevent further slipping. Proceed in the same way with the center and lower kite, and it will be ready for the cover.

The cover tissue should be cut about 1 in. larger all around than the surface to be covered, but turn over about half of this allowance. This will give plenty of looseness to the cover. For the fringe festoons, cut strips of tissue paper, 2½ in. wide, paste ½ in. of one long edge over a string, and cut slits with scissors at intervals of 1 in. along the loose edge. After the fringe has been made, attach it as shown in the illustration. Do not stretch it tightly, but give sufficient looseness to make each length form a graceful curve and keep the sides well balanced.

To bend the bows of the upper and center kites, attach a string from end to end of each bow on the back side of the kite and spring in short brace sticks in the manner usual for tailless kites.

Attach the upper end of the bridle at A. The length of the bridle string is 87 in. and the kite line is attached to it 30 in. from A, leaving the lower part from this point to F, where it is tied to the spine, 57 in. long.

The kite should fly without a tail, but if it dodges too much, attach extra streamers to the ends of the bow sticks of the lower kite, and to the bottom of the spine.

If good combinations of colors are used a very beautiful kite will be the result, and one that will fly well.

Simple Experiment in Electro-magnetism

The following simple experiment, which may be easily performed, will serve to prove the theory that there

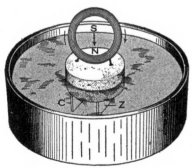

A Small Coil of Wire Mounted on a Cork Floating in Dilute Sulphuric Acid

is a magnetic field produced about a conductor carrying a current, and that there is a definite relation between the direction of the current in the conductor and the direction, or polarity, of the magnetic field produced by the current. The current in the experiment is to be produced by a battery consisting of a small copper and zinc plate fastened to the under side of a large flat cork, as shown in the sketch, the whole being placed in a glass or rubber vessel partly filled with diluted sulphuric acid. A small coil of wire is formed and mounted on top of the cork, and its terminals are connected to the copper and zinc plates. The electromotive force generated will cause a current to circulate through the coil from the copper plate to the zinc plate. If the poles of a permanent magnet be presented in turn to the same side of the coil it will be found that there is a force of attraction between one pole of the permanent magnet and the coil, and a force of repulsion between the other pole and the coil. If the same operation be performed on the opposite side of the coil, it will be found that the force between the poles of the magnet and the coil are just the reverse of what they were in the first case; that is, the pole that attracted the coil in the first case will now repel it, and the one that repelled it, will now attract it. Applying one of the fundamental laws of magnetism—like poles attract and unlike repel each other—it can be readily seen that the two sides of the coil are of opposite magnetic polarity.

If the direction of the current around the coil be changed, the action between the coil and the magnet will be opposite to what it was originally, and if the plates be placed in clean water, there will be no current and no attraction or repulsion between the coil and the poles of the magnet.

Double Lock for a Shed

Four boys using the same shed as their workshop wished to lock it so that any one of them could enter alone. Usually only two keys are supplied with a lock, so two locks were purchased and applied to the staples as shown. Each boy was provided with a key and could enter at his pleasure.—Contributed by George Alfred Moore, Versailles, O.

Ferrules for Tool Handles

Discarded metal caps from broken gas-mantle holders should be saved, as they will come in handy for several purposes, such as ferrules on wood handles, and the like. The wire screen is removed from the end, and the cap is fastened to the handle with a nail or screw.—Contributed by James M. Kane, Doylestown, Pa.

Mallet Made from Wagon-Wheel Felly and Spoke

When in need of a mallet and if an old broken and discarded wagon wheel is at hand, one can be made quickly as

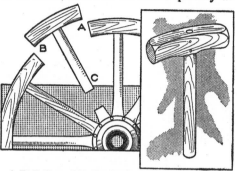

A Well-Shaped Mallet Made from a Section of a Wagon-Wheel Felly and Spoke

follows: Cut through the rim at A and B, and through the spoke at any distance desired, as at C, for instance. The spoke is dressed into the shape of a handle and sandpapered smooth. The section of the felly is used as head and is shaped properly and fastened to the handle with two nails.—Contributed by Mark Gluckman, Jersey City, New Jersey.

A Mystery Sounding Glass

Procure a thin, tapering drinking glass, a piece of thin, black thread, about 2 ft. long, and a long lead pencil. Cut a small groove around the pencil near one end. Make a slip noose in each end of the thread and slip one into the notch and place the thin glass in the other with the thread near the top. When the pencil is revolved slowly the thread will be wound on it slightly and it will slip back with a jerk that produces a ring in the glass. This may be kept up indefinitely. The movement necessary is so small that it is imperceptible. The glass can be made to answer questions by two rings for "yes" and one ring for "no."

⊄A lighted match held to the outside of a fish-pole joint causes an expansion of the outer ferrule and allows the pole to be readily pulled apart.

Repairing a Broken Canoe Paddle

While paddling a rented canoe one day the paddle struck a rock and snapped in two a little below the center of the handle. The boatman laughed at the idea of trying to fix it, but after paying his price for the paddle I decided to try mending it. The barrel of an old bicycle pump was procured and I found that it fitted over the paddle at the break a trifle loosely. It was pushed on the handle out of the way. Then with a No. 8 bit I bored a hole 8 in. deep in the end of each broken part. Into these holes, which formed one cavity when the broken ends were brought together, was forced and glued a tight-fitting 16-in. dowel pin. The outside of the handle was then wrapped with tape for about 10 in. each side of the break, and the pump barrel was forced down over this tape until it completely and firmly enveloped the broken ends.—Contributed by Clarence G. Meyers, Waterloo, Iowa.

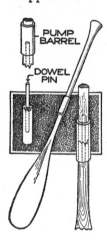

Tightening Lever for Tennis Nets

Tennis nets are always sagging and to keep them at the proper height requires considerable attention, especially so where the posts are not solidly set in the ground. A very effective net

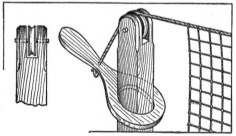

The Upper Rope on a Tennis Net Held Taut with a Lever on the Post

tightener, and one that is easy to make, is the lever shown in the illustration. One end of a piece of hardwood board is shaped into a handle the other end being left large. In the latter a hole is cut to fit loosely over the post for the net. The upper end of the post is notched and a sheave pulley is placed in it so that the groove will be in line with the net. The upper rope on the net is run over the pulley and is attached to the lever handle. A downward pressure on the handle draws the rope taut and locks it on the post. It is easily removed from the post and can be left attached to the rope and rolled up in the net when not in use.

A Desk Watch Holder

A watch holder for the desk is a great convenience for the busy worker, and many calendar devices are sold for this purpose, yet they are no more efficient than the one illustrated, which can be made from an ordinary spindle desk file. If the wire is too long it can be cut off and the bend made in it to form a hook for the watch ring.

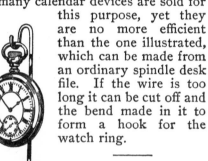

Cleaning Silverware

To clean silverware, or anything made of the precious metals, such as jewelry, etc., is very simple with the following method: Place a piece of zinc in a cup, dish, or any glazed ware; put in the articles to be cleaned, and pour over them a hot solution of water and carbonate of soda—washing soda—in proportions of one tablespoonful of soda to ½ gal. of water. This is a solution and method used by many jewelers for cleaning pins, rings, chains, and many other small articles made in gold and silver.

⫐A machine should never be stopped in the midst of a fine cut.

An Eight-Pointed Star Kite

By CHARLES M. MILLER

NEARLY every boy can make kites of the several common varieties without special directions. For the boy who wants a kite that is not like those every other boy makes, an eight-pointed star kite, decorated in an original and interesting manner, in various colors, is well worth while, even if it requires more careful work, and extra time. The star kite shown in Fig. 1 is simple in construction, and if carefully made, will fly to a great height. It is balanced by streamers instead of the common type of kite tail. Any regular-shaped kite should be laid out accurately, as otherwise the error appears very prominent, and unbalances the poise of the kite.

The frame for this star kite is made of four sticks, joined, as indicated in Fig. 5, with strings running from one corner to the second corner beyond, as from A to C, from C to E, etc. A little notching of each pair of sticks lessens the thickness of the sticks at the center crossing, and strengthens the frame. The sticks are ¼ by ½ in. by 4 ft. long. They are set at right angles to each other in pairs, and lashed together with cord, and also held by a ¾-in. brad at the center. The strings that form the sides of the squares, A to G, and B to H, must be equal in length when tied. The points where the strings forming the squares cross each other and the sticks are also tied.

The first cover, which is put on with paste, laying it out on a smooth floor or table as usual in kite making, is plain light-colored paper. The darker decorations are pasted onto this. The outside edges of the cover are turned over the string outline, and pasted down. The colors may be in many combinations, as red and white, purple and gold, green and white, etc. Brilliant and contrasting colors are best. The decoration may proceed from the center out,

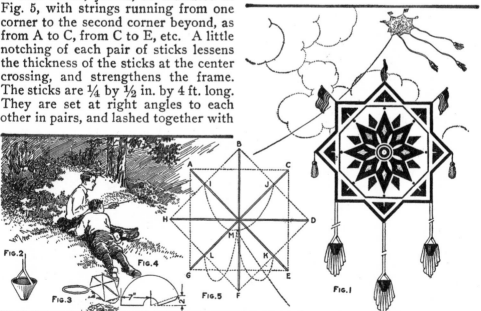

The Boy Who Makes a Star Kite of This Type will Have a Construction Different from the Common Run of Kites, Especially If He Decorates It in an Attractive Manner

or the reverse. The outside edge in the design shown has a 1½-in. black stripe. The figures are black. The next octagonal black line binds the design together. The points of the star are dark blue, with a gilt stripe on each. The center design is done in black, dark blue, and gilt.

The flags are tied on, and the tassels are easily made of cord. The outside streamers are at least 6 ft. long, and balanced carefully. Ribbons, or dark-colored lining cambric, are used for them. The funnel-shaped ends balance the kite. They are shown in detail in Figs. 2, 3, and 4, and have 1-in. open-

ings at the bottom, through which the air passes, causing a pull that steadies the kite. They are of dark blue, and the cloth fringe is of light blue. A thin reed, or fine wire, is used for the hoop which stiffens the top. Heavy wrapping or cover paper is used to cover the hoops. It is cut as shown in Fig. 4 and rolled into shape.

A four-string bridle is fastened to the frame at I, J, K, and L, as shown. The upper strings are each 18 in., and the lower ones 32 in. long, to the point where they come together, and must be adjusted after the kite line is fastened at M.

Second Handle on Hoe or Rake Saves Stooping

Anyone who has used a hoe or rake for days at a time will appreciate the labor saved by the attachment for the

Much of the Tiring Labor in Using a Hoe or Rake is Overcome by This Simple Homemade Attachment

handle shown in the illustration. It is adjustable to various-sized persons by means of the holes at the front end of the horizontal piece. The two parts are each made of strips joined at the middle portions, and arranged to clamp on the handle of the hoe or rake. In hoeing around shrubs and large plants, the handle may be set to one side.—A. S. Thomas, Amherstburg, Ont., Canada.

Photo-Copying Lens Increases Angle of Camera

Trying to take some indoor pictures, I found the angle of my ordinary lens was insufficient to "get in" the various objects I desired. Not having a wide-angle lens, I decreased the focal length of the lens by using a copying attachment. The results were quite pleasing, and while there is some distortion and less of the plate is covered than usual, there is a remarkable increase in the angle of view. To obtain definition, it is necessary to stop the lens down, but the pictures are very clear.—Samuel L. Pickett, Denver, Colo.

Belt for Sprocket Drive Made of Brass Strips

Being unable to purchase a small driving chain for sprockets made by cutting out every other tooth in gears taken from a clockwork, I used a brass strip, properly punched, and found it satisfactory. The

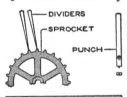

strip was .005 in. thick and the points where the holes were to be punched out were indicated by dividers set from the gears. I made a punch from a nail, leaving a small center on it as shown, and grinding the end to an oblong shape. I used a piece of sheet lead as a die, on which to punch the strips. The marks made by the dividers provided spots on which to set the center of the punch, making the result quite accurate.—Edward M. Davis, Philadelphia, Pa.

Rain Alarm with Drop-of-Water Contact

An annunciating device, which awakens a person sleeping in a room with the window open and warns him that it is raining, so that he may close the window, is an interesting bit of electrical construction. On the outside of the house, as detailed, is a funnel fixed to the wall. At its small end, two separate wires have their terminals. The wires enter the room at the frame of the window, and connect to an electric bell, and a dry cell. A drop of water entering the funnel, flows down to the small end, falling on the terminals of the wires, and acting as a conductor, completes the circuit, ringing the bell. A

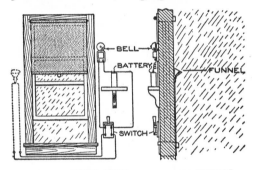

INSIDE VIEW   OUTSIDE
A Drop of Rain Water Completes the Bell Circuit, Thus Giving Warning of the Rain

switch inside cuts out the circuit, stopping the bell's ringing.—John M. Chabot, Lauzon, Quebec, Can.

Coaster Steering Gear Made from Cream-Freezer Drive

In rebuilding a wagon into an automobile coaster, I used the driving rod and gears from an old ice-cream freezer, and found that it worked so well that perhaps other boys might be interested

in the job. The front of the coaster was covered with a hood, and the steering wheel was set back of it, as shown. The center rod of the freezer was used

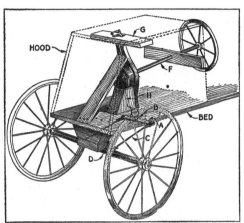

A Steering Rig That Works Almost Like That on an Automobile was Made Out of the Driving Parts of an Old Ice-Cream Freezer

for the steering post F, and an old rubber-tired wheel was made into a steering wheel. The casting from the top of the freezer, with the gears in it, and the rod on which the turning crank was fastened were set on a block, H, and braced, as shown. The shaft where the crank was fastened, at B, was set through the wagon bed. A crosspiece of iron, A, was wired to the axle D with wires C. A heavy block was used for a turntable. The top end of the casting was fastened to the hood with a brace, G, and the block H steadies the rigging also.—L. Chester Bryant, El Dorado, Ark.

Pad for Glass Vessels Made of Corks

In the kitchen, shop, laboratory, and other places where glass or other fragile dishes or vessels are used, a convenient pad on which to rest them can be made by stringing corks on a strong cord or wire in the form of a ring. Several rings of corks may be used to make a mat, or rings slightly larger in diameter than the bottle or vessel may be made for certain sizes of containers. If desired, the corks may be cut to fit closely on the radial joints, making the resulting ring more secure.

A Shaving Lamp and Mirror for the Camp

To make shaving possible in camp at night, or with little daylight, a small mirror was provided with an electric flash light. The mirror was set to swing free, in a wooden support. The light was fastened slightly above and behind the mirror, and swings at its base, so that it can be tipped upward or downward, throwing the light correspondingly. A piece of wood, 1¼ by 3½ in., and as long as the mirror frame is wide, serves as a base. The arms will hold the mirror far enough in front of the lamp to allow room in which to swing. The body of the lamp is set on a block, and held between two wooden pieces, into which a band of iron was set near the top. The uprights move in an arc, pivoting at their lower fastening, on screws.—C. L. Meller, Fargo, N. D.

Automatic Electric Light on Talking-Machine Cabinet

In many homes the phonograph is placed where little light is available in changing the records, setting the needle, etc. An electric light which is lighted only while the cover of the phonograph is raised, is well worth installing. A metal arm, A, supports the open cover of the cabinet. When the cover is closed, this arm passes through a slot and takes the position shown by the dotted line. A

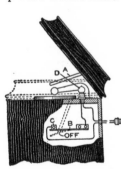

strip of spring brass, B, is fastened to the inside wall of the cabinet, in the path of the arm, so that it will be pushed down to the off position, as indicated. When the arm releases the strip B, the latter presses against the contact C. A small electric lamp, D, is set in the corner, and electrical connection made to it through B and C, the plug connections passing through the back of the cabinet. When the cover is down, the electric circuit is open, and the moment it is raised, connection is made at C, and the lamp lights. The backs of most phonograph cabinets may be removed easily to make these changes.—M. C. Ball, Kansas City, Mo.

Device for Suspending Parcels from Overhead Hooks

To hang small sacks or other articles out of reach overhead, so that they may

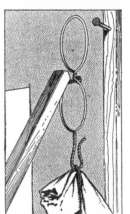

be easily taken down, I use a double-eye hook which I made of wire. A single piece of wire is used, and twisted into two loops, as shown, and then formed into a twisted hook. I use a pole with a nail, hooking it into the lower loop, to raise the parcel; this leaves the upper loop free to be hooked on the nail above.—E. B. Warren, Garnet, Mont.

Steel Wool as Aluminum-Ware Cleaner

It takes little trouble to keep aluminum pots and pans shining if they are cleaned frequently with steel wool, water, and a nonalkaline soap. Use a very fine grade of the wool, and give the utensils a few rubs frequently, rather than attempt to clean them only occasionally, when much soiled.—L. P. Langan, Denver, Colo.

A Submarine Camera
by Charles I. Reid

SUBMARINE photography should have great attractions for amateur photographers who have access to lakes, ponds, and other clear waters. While more careful work is demanded than in ordinary photography, the method of obtaining good results is not difficult, and the necessary equipment may be provided by constructing the device shown in the illustration. Submarine pictures can be taken in a considerable depth of water, providing it is reasonably free from foreign matter. This is a fascinating field of photography, and many pictures of educational and scientific value remain to be made of under-water life. The illustration shows the detailed construction of the camera chamber, and the method of suspending it from a bridge, or other place convenient to the body of water. Reproduced in the oval panel is a photograph of fish near baited hooks, on a fishline. The original was made from a negative exposed by the use of the camera chamber described.

The problem of making photographic exposures under water involves the provision of a strong water and pressure-proof container for the camera, a means for controlling the shutter, and a suitable opening in the container through which the exposures may be made. The arrangement described combines these features in a simple manner, and by the use of materials that can be obtained without difficulty. It was made for a camera taking 4 by 5-in. pictures, and the dimensions given are for a container for this size. The dimensions may be varied to adapt the device to various cameras, within reasonable limits. A 9-in. steel pipe was used for the chamber, and its ends were fitted with pipe caps. A heavy piece of plate glass was fitted into the forward cap, which was cut into the shape of a ring, to provide the exposure opening. The general arrange-ment of the camera in the chamber is shown in the sectional view, Fig. 1, as seen from the shutter end. The electrical device, by which the shutter is controlled, is shown in this view, and in Fig. 2 it is shown in detail.

The chamber was made as follows: A section of 9-in. steel pipe was cut to a length of 11½ in. and threaded on the ends to fit pipe caps. The forward pipe cap was chucked up in a lathe and the center portion cut away, to provide an exposure opening and a shoulder at the rim, on which the plate-glass window rests. A graphite paint was applied to the rim, then the glass was bedded solidly in it, and a rubber gasket was fitted to the joint, making it waterproof when the cap was drawn up tightly. The chamber assembled and in detail is shown in the illustration.

Holes were bored into the top of the chamber, and eyebolts were fitted into them. Between the eyebolts a hole was bored and fitted with a water-tight collar, through which the wires leading to the shutter-control device pass. The chamber is supported by the wires, which are fixed to the eyebolts and secured at the base of operations by the photographer.

A support for the camera was provided by bending a strip of ⅛ by 1-in. band iron to the shape indicated in Fig. 1, at A, and riveting it to the bottom of the chamber. Its upper surface is flat and was bored and threaded to fit the tripod thumbscrew B, on the lower surface of the camera. The camera is arranged on the support and clamped into place firmly by the thumb nut, as it might be on a tripod. The adjustment of the camera in the chamber is done from the rear, and the space beneath the thumbscrew should be large enough to make access easy. A camera of the size indicated, when fitted with its lens centering on the center of the window, will be raised sufficiently for convenience in clamping it. The threads on the back cap must fit snugly and no paint must be used on them. Hard oil, or vaseline, may be applied to insure a water-tight joint that permits easy removal of the cap.

The making and adjustment of the electrical shutter device requires care, but its operation is simple. An electromagnet, of the type used on doorbells, was fixed to the front of the camera, above the shutter, as shown in Fig. 1, and in detail in Fig. 2. It is actuated by current from two dry cells. The latter are kept in a convenient carrier at the base of operations, and are connected to the magnet by a single strand of double, waterproof wire. This is spread as it reaches the chamber and fastened to the two eyebolts in the top. The ends of the wires are conducted through the water-tight center opening between the eyebolts, and attached to the magnet. The release lever is fitted to a steel hook, pivoted at its upper end with a small nail, C, Fig. 2. A rubber band is fixed to the lower edge of the shutter lever and its other end is attached to the front of the camera. When the current is permitted to flow into the magnet by pressing a contact key, in the hand of the operator, the steel hook is drawn from the release lever, and the rubber band draws the lever down, making an exposure.

The double-wire cable carries the current as well as holds the chamber suspended in the water. The wire should be about 25 ft. long, and, in transporting the outfit, or when only partly used, is coiled. The chamber should be completed for picture-taking operations by giving it a coat of dull, black waterproof paint, both inside and outside. This will prevent rusting and also serves to make the object inconspicuous when in the water. It is important that the interior be painted in this manner, because reflections of light within the chamber may cause difficulty in obtaining satisfactory results. When the paint is thoroughly dry, the device may be tested for leakage and assembled ready for a test before making an actual trial in the water. The camera is fitted into the chamber so that it centers on the center of the plate-glass window, and is clamped into place. If the electrical device operates satisfactorily the plate may be inserted, the plate-holder slide withdrawn, the back cap replaced securely, and the outfit lowered into the water. It should be watched carefully until it reaches the proper depth, for, if it is permitted to touch the bottom, the sediment stirred up must be given time to settle before an exposure is made. The forward end of the chamber should be marked on its upper edge with a streak of white paint, to aid in identifying it at considerable depth in the water. This is important, since the operator must shift the chamber carefully until the window faces the objects to be photographed. When the chamber is in position, the contact key is pressed, and the exposure is made.

The time of exposure for under-water photography depends on the clearness of the water, the depth at which the pictures are to be taken, and the light conditions on the surface. A bright day is, of course, desirable for this class of photography. A safe approximation, on a sunny day, in clear water, and with the chamber lowered to a depth of 20 ft., is $\frac{1}{25}$ sec. at the F 8 stop. The fastest plates or films obtainable should be used for this work, making possible a fairly rapid shutter speed. This tends to overcome the movement of the subject and possible movement of the camera.

The camera should be focused while in the chamber in order that the plate glass may not disturb the focus. The glass usually changes the focal length of the lens slightly, hence this precaution must be taken. The camera should be focused in the chamber for a distance of 10 ft., as this is the average at which under-water photographs will be taken ordinarily.

When attempting under-water photography in cloudy waters, or at a considerable depth, the necessary illumination may be provided by a charge of flash-light powder. For this purpose another submarine chamber, similar to

THREADED

PLATE GLASS

IRON CAP THREADED

IRON PIPE 12

A B FIG.1

MAGNET

C

STEEL HOOK

RELEASE LEVER

RUBBER BAND

FIG.2

WATER LINE

Photographing Subjects under Water Is a Fascinating Diversion, and Each Exposure Has an Element of Mystery in the Uncertainty of the Result. The Photograph Reproduced in the Oval was Taken with the Outfit Shown. The Construction of the Chamber is Shown at the Middle. Fig. 1 Shows a Sectional Interior View, and Fig. 2, a Detail of the Electrical Shutter Release

that used for the camera, should be provided, with a plate glass, ½ in. thick, and a valve fitted into the top of the chamber, and opening outward, so that the gas may escape. Fifteen grains of powder will suffice, and this should be set off by a small electrical fuse connected to the current supply.

Every pond, lake, and river abounds in interesting and instructive subjects for submarine photography. Along the coast of Florida, and at many points along the Pacific coast, are waters of such clearness that pictures

may be taken at a depth of nearly a hundred feet, without the use of artificial illumination. These localities abound in objects under water of great interest, such as shipwrecks. The fascinating art of taking pictures under water does not make it necessary for one to go to these places, for subjects are easily available. Wnenever the submarine chamber is raised from the water there is an element of mystery involved, regarding what may be recorded on the plate or film, and this is an attractive feature of the diversion.

The Magic of Numbers

By JAMES L. LANYON

That there are a great many magic squares; that the numbers in these squares are arranged according to a definite system; that squares with very remarkable properties are easily constructed, are facts not generally known.

Consider the magic square A of 16 numbers. Add up any four numbers straight across, up and down, or diagonally—10 ways in all—and the sum in each case will be 34. But that is not all: Take the four numbers in any one quarter of the square, as for example, 15, 10, 4, and 5, and the sum will be 34; or take the four central numbers, or the four corner numbers, and the result will be the same. But even this does not exhaust the magic of the square. Add any four numbers arranged symmetrically around the center, as 3, 10, 8, and 13, or 10, 4, 7, and 13, and the result will also be 34. In fact, it is really not necessary to have them arranged symmetrically, because it will be found that four numbers arranged as are 6, 10, 11, and 7, or 1, 4, 16, and 13 will produce the same magic number of 34.

There are two other combinations of the 16 numbers that will give the same result. They are shown at B and C. In fact the second one, B, not only exhibits some of the former combinations, but also includes such sets of four as 14, 5, 3, and 12, or 15, 8, 2, and

9, which places to the credit of this square numerous combinations. Such special features as this simply add another element of mystery and interest. Thus, while the square B has these two combinations exclusively to its credit, the first, A, and the third, C, have such special arrangements as 5, 16, 1, and 12, or 15, 6, 11, and 2. Also 10, 3, 5, and 16, or 4, 5, 14, and 11, making the total number of such combinations for the first square 34.

Magic squares of 25 numbers also have remarkable properties. Examine the square D and note the many possible combinations graphically set forth in the small diagrams. Not only do any five numbers in a row or along a diagonal make 65, but almost any four arranged around the center, with the center number 13 added, will give the same result.

This square is a good example by which to illustrate one of the methods of construction of these interesting devices. Thus, place 1 in the middle square of the top row, and then write the numbers down consecutively, always working in the direction of the arrows as indicated. When any number falls outside, as number 2 does at the start, drop down to the extreme square in the next row and insert the number there, as was done in this case. It will be observed that 4 falls outside, and so it is moved to the proper square as suggested, which will be at the extreme left of the next row above. Continuing, it is found that at 6 it is neces-

Although they do not contain quite so many combinations, the three magic squares shown at G all add up to this

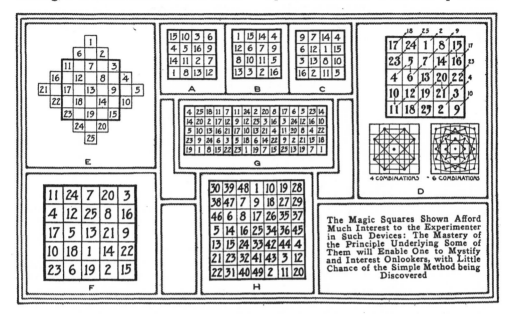

E

A

B

C

18	25	2	9
17	24	1	8	15	17
23	5	7	14	16	23
4	6	13	20	22	4
10	12	19	21	3	10
11	18	25	2	9

G

4 COMBINATIONS   6 COMBINATIONS

D

11	24	7	20	3
4	12	25	8	16
17	5	13	21	9
10	18	1	14	22
23	6	19	2	15

F

30	39	48	1	10	19	28
38	47	7	9	18	27	29
46	6	8	17	26	35	37
5	14	16	25	34	36	45
13	15	24	33	42	44	4
21	23	32	41	43	3	12
22	31	40	49	2	11	20

H

The Magic Squares Shown Afford Much Interest to the Experimenter in Such Devices: The Mastery of the Principle Underlying Some of Them will Enable One to Mystify and Interest Onlookers, with Little Chance of the Simple Method being Discovered

sary to drop down one square and continue in the direction of the arrows. At 9 it is necessary to drop down to the proper extreme square as shown. The next number, 10, must again be provided for at the square on the left of the next higher row. The square ahead being already filled, 11 is placed below; after this there is "clear sailing" for a time. In this manner magic squares with seven or nine numbers to the side may be made easily. When puzzles and catch problems are under discussion, it is always mystifying to take one's pencil and quickly make out a magic square according to this easily remembered method. The small diagrams at D suggest some of the combinations.

Another method of constructing a square of 25 numbers diagonally is shown at E. Place the outside numbers in the open spaces at the opposite side of the square, maintaining the same triangular relation, which results in the arrangement shown at F. While this combination is entirely different from the previous one, it exhibits the same mysterious properties.

same magic number of 65, straight across, diagonally, and many other ways. A square with seven numbers to the side, worked out according to the first method described, is illustrated at H. The magic number here is 175. Since the general principle is similar to that involved in the squares described in detail, the working out of the numerous combinations of the squares shown at G and H will be left to the interested experimenter.

Muffling the Ticking of a Watch or Clock

When a watch is used on a table in the sick room, the ticking may be eliminated by placing an ordinary tumbler over the watch. The face may be seen readily. A large glass vessel, or transparent jar, may be used in the same way to cover a small clock.—L. E. Turner, Jamaica, N. Y.

⁋To letter on prepared cloth, use colors ground in japan and thin the mixture with gasoline to the consistency of cream.

A Simple Cipher Code

Adapted for Use in Private Correspondence

By CAPT. W. H. WALDRON, U. S. ARMY

HAVE you ever needed a secret code in which to couch the contents of a message intended for the eyes of one person alone? If you have, you will remember the difficulties that were experienced in making up the code and enciphering your letter. Here is a cipher code that may be mastered

Fig.1   Fig.2   Fig.3

The Cipher Code Illustrated in These Diagrams may be Adapted for Wide Uses by the Substitution of Appropriate Key Words for Those Shown

in a few minutes; one that is most difficult to decipher by any person other than those having the key words, and that is very simple when once understood.

It is commonly known as the "Play Fair" code and is in use in some of the foreign military services. It is a substitutive cipher which operates with one or more key words, two letters in the code being substituted for each two letters in the text of the message. In preparing the cipher code by this method the key words are selected by the correspondents and their location in the cipher square mutually agreed upon. A large square divided into 25 smaller squares is drawn, as shown in Fig. 1, and the letters of the key words entered into their proper spaces, the remaining spaces being filled by other letters of the alphabet. The key words must not contain duplicate letters. The letters I and J are considered as one and entered in the same space, the letter I being invariably used in enciphering.

Suppose that the two words "grant" and "field" have been selected for the key, the same to be entered respectively in the spaces on the first and third horizontal lines of the square. Then the basis of the construction would be as indicated in Fig. 1. Now fill in the remaining fifteen spaces of the square with other letters of the alphabet, beginning at the blank space at the left of the second line, entering the letters in rotation and not using any letter of the key words. The completed cipher would then appear as shown in Fig. 2.

The text of the message to be sent is then divided into groups of two letters each and the equivalent substituted for each pair. Where two like letters fall in the same pair the letter X is inserted between them and when the message is deciphered this additional letter is disregarded. If one letter is left over after the last pair, simply add an X to it and make a pair.

Suppose it is desired to send this message in the cipher: "Will you meet me as agreed." Having three pairs of the same letter, it will be necessary to break them up by placing the letter X between them. The message will then be paired off as follows:

WI LX LY OU ME XE TM EA SA GR EX ED

The message may now be enciphered, after considering three simple rules for guidance: Every pair of letters in the square must be either in the same vertical line; in the same horizontal line; or at the diagonally opposite corners of a rectangle formed by the smaller squares within the large square.

In the first case, R and P are in the same vertical line (the second), and the next letter below, in each case, is substituted for R and P, which are C and W. If the pair consists of K and Y (fourth vertical), substitute L for K and go to the first horizontal line (fourth vertical) for Y, substituting N for Y. In the second case B and H are in the same horizontal line (the second), and thus substitute the next letters to the right, which are C and K. If the pair consists of P and U (fourth horizontal), substitute Q for P and then go back to the first vertical line (fourth horizontal) and substitute O for U. In the third case, R and S are at the opposite corners of a rectangle. Each letter of the pair is substituted by

116

the letter in the other corner of the rectangle on the same horizontal line with it. Then R would be represented by N, and S would be represented by P. To illustrate further, NE would be represented by AL; BZ would be represented by MV; TP by RU.

The message may now be enciphered, applying the rules:

WI LX LY OU ME XE TM EA SA GR EX ED
RP EY SN PO HD AQ MD QH QN RA QA LF

In sending this message, to make it more difficult for the inquisitive cipher expert, divide the substituted letters into words of five each and give him the added task of determining whether the cipher used is the transposition or the substitution method. The message ready to hand to the telegrapher would read:

RPEYS NPOHD AQMDQ HQNRA QALFX

In deciphering a message the method is reversed. Take the message as received, divide the letters into pairs, and disregard the final X, which was put in to make a five-letter word. Then apply the key reversed. Practice it on the above message to get the system with respect to letters occurring at the end of the lines. Where the letters of a pair are in the same vertical line, substitute for each the letter above; where they are in the same horizontal line, substitute the letter to the left; where they are in the corners of a rectangle, substitute the letters at the opposite corners on the same horizontal line. To test the understanding of the system, the message given in Fig. 3, with the key words "chair" in the first horizontal line and "optun" in the fourth line, may be deciphered. The message to be deciphered is as follows:

FQVUO IRTEF HRWDG APARQ TMMZM RBFVU
PICXM TRMXM AGEPA DONFC BAXAX.

Cheese Grater and Ash Tray Made from a Tin Can

Being in need of a cheese grater and finding it inconvenient to go many miles to town, I constructed a satisfactory makeshift. I took a heavily tinned can and cut it in two, as shown in the sketch. By punching holes through it from the inside a practical grater resulted. From the remaining half of the can I made an ash tray, as shown at the right of the sketch.

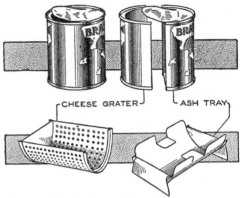

Necessity Resulted in the Making of a Cheese Grater and Ash Tray from a Tin Can

The semicircular ends were bent over to form a rest, and by cutting portions at the sides and bending them in, a convenient rest for a pipe or cigar was afforded.—Gus Hansen, Peachland, B. C., Canada.

An Improvised Typewriter Desk

Travelers and others who carry typewriters on their journeys frequently find it inconvenient to use the tables provided because they are usually too high for typewriters. A method of overcoming this difficulty is to withdraw the drawer from the table and invert it in the slide as shown. The type-

Remove the Drawer and Replace It Inverted, to Provide a Convenient Rest for the Typewriter

writer may then be placed upon the bottom of the drawer and will be considerably lower than if placed upon the table top.

An Inexpensive Imitation Fire

Window decorations may frequently be made attractive by the use of an imitation fire in a stove or a fireplace, when an indoor setting is on display. To produce such an effect, p u t an electric fan below the place at which the imitation fire is to be arranged and run an electric-light cord, with a red globe attached to it, to the center of the "flame." Cut a number of strips of Indian red tissue paper and fix the lower ends of these to form a circular mass above the globe. When the light is turned on permit the fan to direct a stream of air against the tissue-paper ribbons, forcing them upward to appear like tongues of flame. The sketch shows this method applied to a heater. The fan is placed in the ash box and the electric light is conducted through the grate.

Jardinière Made of Metal-Lamp Body

Some of the metal bodies of old lamps, and they are usually brass, are of s u c h ample s i z e a n d so neatly embossed that they can be readily used as jardinières i n - stead of being handed to t h e junk man. It is only necessary to remove the lamp part, and set the plant pot into the bowl of t h e stand, as shown in the sketch.—H. N. Wolfe, Chicago, Ill.

Replacing a Broken Coffeepot Knob

A knob was broken from the lid of a coffeepot, which was valued by reason of its associations. Attempts to fasten the broken portions together were unsatisfactory, and the rough surface, where the earthenware was broken, was ground smooth and fitted with a carefully shaped wooden knob. A hole was drilled through the center of the lid with an old file and the wooden knob bolted into place. When stained appropriately, the repair was quite satisfactory.

Homemade Magnesium Printer

A convenient homemade printing device consists of a smooth board, A, 2 ft. long and 1 ft. wide, and an upright, B, which is 1 ft. square. Bore a hole in the center of the upright for the small tin holder E, to carry the magnesium ribbon, made by folding a piece of tin to fit it. Small pieces of wood, CC, are nailed across the board

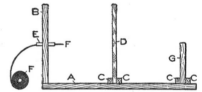

Length of Magnesium Ribbon Burned Determines the Time of Exposure

to hold the ground glass D and the printing frame G. The ground glass is 10 in. from the upright, and the printing frame is 10 in. from the ground glass. The latter is 1 ft. square and is used to diffuse the light from the magnesium ribbon F, which may be purchased from any dealer in photographic supplies. The length of exposure varies according to the length of ribbon which is permitted to burn. This should be tested out carefully before making exposures. — Gustave Straub, Albany, N. Y.

❡In sandpapering a varnished surface between coats, especial care should be taken to avoid rubbing through at sharp edges.

Decorative Toys and Boxes

Made at Home
By Bonnie E. Snow

HOMEMADE toys and gifts, as well as the "treasure boxes" in which they are contained, have an added interest both to the one making and the one receiving them. The holiday season makes this work especially attractive, which affords opportunity for individuality in construction and design limited only by the skill of the worker. The decorated toys and the box described in detail in this article are suggestive only, and may be adapted to a large variety of forms and designs. The gorgeously colored parrot and the gayly caparisoned rider and horse suggest a host of bird and animal forms, those having possibilities for attractive coloring being most desirable. The decorated box shown in Fig. 7 may be adapted as a gift box, to be used where its decoration may be seen, in the nursery, for example, and may be made in many forms, in fact as various as cardboard boxes are. Plant, animal, or geometrical forms may be used to work out designs, and appropriate color schemes applied to them. A good plan in determining upon a color scheme is to use the colors of the flower or other motif. If the design is not associated

with objects having varied colors—a geometrical design, for instance—harmonious colors should be chosen. These may be bright and contrasting, as red and green, violet and orange, or subdued in tone.

A design for a horse and rider, brightly colored, is shown in Fig. 1. The form is cut out of thin wood, the color applied, and the figure mounted on the curved wire, weighted at one end, as shown in Fig. 6. The toy adds a touch of color and novelty to a room, when suspended from the corner of the mantel, from a shelf, or other suitable place. Balanced in a striking attitude, forefeet upraised, even grownups can hardly resist tipping the rider to see his mount rear still higher. The parrot shown in Fig. 2 is made similarly, and is weighted at the end of the tail. The point of balance is at the feet, which may be fastened to a trapeze, or be arranged to perch on a convenient place, like that suited to the horse and rider.

The tools and equipment necessary for the making of such toys are simple, and available in most boys' workshops or tool chests. A coping saw, like that shown at A, Fig. 3, is suitable for cut-

The Outlines for the Horse and Rider and the Parrot may be Made by Enlarging These Sketches. The Color Scheme Indicated Is Suggestive Only and may be Varied to Suit Individual Taste

ting the wood. A fretsaw, operated by hand, foot or power, may be used, and such a tool makes this work quite rapid. To use the coping saw to the best advantage, particularly if the work is to be done on a table which must not be marred, a sawing board should be made. In its simplest form, this consists of a board, as shown at B, about $7/8$ in. thick, $3\frac{1}{2}$ in. wide, and 6 in. long, with one end notched. This is clamped to the end of the table, as at D, with a clamp, an iron one of the type shown at C being satisfactory. Another form of sawing table especially useful when it is desired to stand up at the work, is shown at E in detail and clamped in the vise at F. It consists of a notched board, $3\frac{1}{2}$ in. wide, fixed at right angles to a board of similar width, 11 in. long, and braced at the joint with a block about $1\frac{3}{8}$ in. square. In using the coping saw with either of these saw tables, the wood is held down on the support, as shown in Fig. 5, and the saw drawn downward for each cutting stroke, thus tending to hold the board more firmly against the saw table. It is, of course, important that the saw

described may be followed. The wood must be sawed slowly, especially at the beginning of a cut. The operator soon learns the kinks in handling the saw and wood to the best advantage, and can then make rapid progress.

An outline drawing of the form to be cut out of the wood must first be made, to the exact size that the object is to be. There is much satisfaction in working out the form of the animal or other figure, especially for the boy or girl who has the time necessary to do good work. If desired, the figure may be traced from a picture obtained from a book, magazine, or other source. Cut a piece of wood to the size required for the design, and place a sheet of carbon paper over it; or if none is available, rub a sheet of paper with a soft pencil, and use this as a carbon paper, the side covered with the lead being placed next to the wood. The carbon paper and the sheet bearing the design should then be held in place on the wood with thumb tacks, or pins, and the transfer made with a pencil, as shown in Fig. 4. The design should be placed on the wood so that the

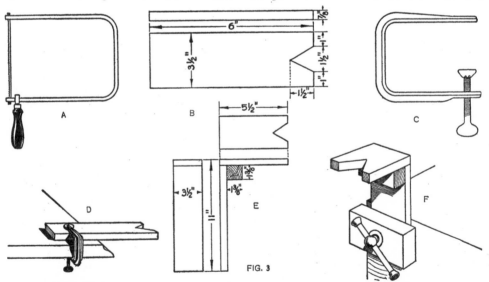

FIG. 3

The Tools Required are Found in Most Boys' Workshops, and a Satisfactory Saw Table may be Made Easily, as Shown in Detail

be inserted in the coping-saw frame with the teeth pointing toward the handle, so that the method of cutting

weaker parts, such as the legs of the horse, will extend with the length instead of across the grain of the wood.

In some instances, where a complicated form is cut out, it is necessary to use wood of several plies, and where

put on first and the other colors applied over it, when dry.

Oil paints may be used, and a var-

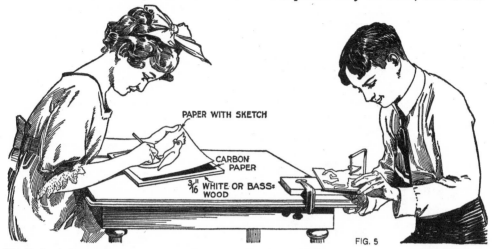

PAPER WITH SKETCH

CARBON PAPER

$\frac{3}{16}$" WHITE OR BASS-WOOD

FIG. 5

The Design is Traced Carefully onto the Wood and Then Cut Out with the Coping Saw, on the Saw Table

this is available it is worth while to use it for all of this work. For smaller objects wood $\frac{3}{16}$ in. thick is suitable, and stock up to $\frac{1}{2}$ in. in thickness may be used. Whitewood, basswood, poplar and other soft, smooth-grained woods are suitable.

When the design has been outlined satisfactorily, place the piece of wood on the saw table with the design on the upper side. Holding the wood down firmly, as shown in Fig. 5, and sawing in the notch of the saw table, cut into the edge slowly. Apply light pressure on the downstroke only, as the upstroke is not intended to cut, and turn the piece to keep the saw on the line and in the notch. It is important that the saw be held vertically so that the edge of the cut-out portion will be square. With proper care and a little practice, the edges may be cut so smoothly that only a light sandpapering will be required to produce a smooth edge. When the figure has been cut out, smooth the edges by trimming them carefully with a sharp knife, if necessary, and sandpaper them lightly to remove sharp corners. A fine sandpaper, about No. $\frac{1}{2}$, is suitable for this purpose. The figure is then ready for painting. The white is

nish or shellac applied over them to give a high grade of work, but this process requires much care, considerable skill, and long drying between coats to prevent "runs" in the colors.

Water-color paint, which can be purchased in powder form at paint stores, mixed with water to the consistency of cream, is a satisfactory coloring material, and is easy to apply. Five cents' worth of each of the colors used— yellow, red, blue, black, and white— will be sufficient for several toys. Mix

FIG. 6

each color in a separate saucer, and use a small water-color brush to apply the paint. In painting the horse and rider, the horse is first painted entirely white, and then the black spots are applied after the color is dry. The

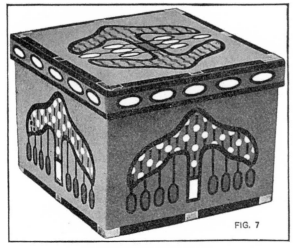

FIG. 7

A Handmade Box Is Interesting in that It Expresses the Individuality of the Maker, Especially When Made as a Gift

rider's coat is painted red; the trousers blue; the hat and leggings buff, as indicated in Fig. 1. Mix a brushful of yellow with a brushful of red, and add about three brushfuls of white. A half brushful of black may be added to dull the color, if desired. The flesh tone for the rider's face is made by mixing a little red with white. When the colors are dry, all edges are outlined with a heavy line of black, not less than 1/8 in. in width. This outline may be evenly applied with the point of the brush.

The method of making the parrot is similar to that described for the horse and rider, and the color scheme is suggested in Fig. 2.

It will be noticed that no attempt is made to secure a lifelike, or realistic, effect in painting these toy shapes. All colors are flat, that is, without light and shade. The toys are really decorative designs, and the maker is at liberty to use any colors desired, whether natural or not.

The horse and rider is balanced on the hind foot, as shown in Fig. 6, by using a lead weight, attached to a 1/16-in. wire, as a counterweight. The wire should be set into the body of the horse, behind the foreleg, to a depth of

3/4 in. The weight of the metal and the curve of the wire should be adjusted to obtain the proper balance. The parrot is balanced in the same way, except that the weight is fixed to the end of the tail, which is curved like the wire.

These and other homemade toys or gifts may be sent or contained appropriately in boxes decorated to match them, as shown in Fig. 7. They may be made complete, or commercial boxes of suitable sizes may be covered and decorated. If good materials are used, such a box makes a pretty and useful gift in itself. The complete process of making a typical box is described for those who prefer to make one of special size. The dimensions given are thus only suggestive, and may be adapted to suit particular needs.

The materials necessary are: cardboard, cover paper, lining paper, bond paper, paste, and water colors. The latter should be of the opaque variety, since white or other light shades may then be used on darker-colored paper.

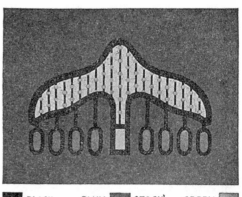

BLACK   PLUM   STOCK   GREEN

YELLOW   FIG. 8   RED

Color Schemes may be Obtained from Flowers and Other Natural Forms, or by Selecting a Combination of Harmonious Shades. The Background Is of Plum-Colored Paper; the Small Circles are Emerald-Green; the Light Area, Yellow, and the Ovals, Orange-Red

A sharp knife, a scissors, a metal-edge ruler, and bookbinder's paste are also needed. Suitable substitutes for the various kinds of paper may usually be obtained in the home, if they are not readily available at local stationery stores or printing establishments.

from the long edges, and then spread a thin layer of paste over the outer surface of one of the sides of the box. Apply the paper to the pasted surface

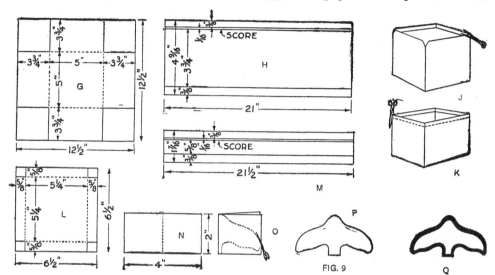

The Various Steps in the Process of Making and Covering a Rectangular Cardboard Box are Shown in Detail. The Method of Making a Pattern for the Design is Shown Below

The box is made as follows: Determine upon the proper size and select materials to carry out the design. An appropriate combination of colors and materials is suggested in Fig. 8. Cut out a square of the cardboard, having sides 12½ in. long, as shown in Fig. 9, at G, then mark it as indicated and cut on the full line to remove the square corners. Crease it on the dotted lines and fold to form a box. To hold the cardboard in box shape, strips of bond paper—ordinary writing paper—are cut, 3¾ in. long and 1 in. wide, then creased along their centers and pasted to the corners. The paste should be applied to the paper strip first, then on the corners of the box. Apply the piece of paper over the corner of the box on the outside, pressing it to make a snug fit. Repeat this operation on the other corners.

Lay off the dimensions given at H on the selected color of cover paper, which in this instance is plum, and score the lines indicated. Spread paste smoothly over the surface of the plum paper, between the lines drawn ⅜ in.

and press it down, rubbing gently out from the center to remove air bubbles. Fold the ⅜-in. laps at the top and bottom over the upper edge of the box and around the lower corner. Repeat this process, covering the four sides. To form a smooth fold at the corners, it is best to miter the paper as shown at J and K, before pasting it down. Then paste a square of the same paper 4⅞ in. wide on the bottom of the box, taking care to match the edges evenly all around.

Line the box with a strip of lining paper, 20 in. long and 4 in. wide. Try the lining by folding it into the box so that its upper edge is about ⅛ in. from the edge of the box, and crease it carefully into the corners. Remove it, apply paste, and press it well into the corners when pasting it down. Paste a square of the same paper, 4⅞ in. wide, in the bottom of the box.

The cover is made by the same process as the main portion of the box. The dimensions of the cardboard are shown at L, and the covering at M. It should be observed that the cover is slightly wider than the box, so that it will fit easily.

When the box is thoroughly dry, it is ready to receive the decorations on the top and sides. The design may be adapted from a leaf, flower, or similar form, as well as from geometric or animal forms. The horse and rider, the parrot, and the animals shown in the headpiece of this article are all suggestive of animal forms that are available.

The design shown on the box in Fig. 7 was adapted from a flower form, two of the units being joined for the decoration on the top. To obtain a pattern for the design, fold a piece of paper, 2 by 4 in., as shown at N, and outline one-half of a leaf, flower or similar motif. Cut the folded paper as at O, and a pattern similar to that shown at P results. Trace around this to place the figure on the box. Outline the figure with black, about ⅛ in. wide, as at Q. The oval figures, suggestive of small pods on a flower, are also outlined in black, and joined to the main portion of the design by a black line. Thinner black lines are drawn vertically across the form, and small circles placed along them at intervals. The color scheme is shown in Fig. 8. The light background is yellow, the small circles are emerald-green, and the ovals red-orange. The rim of the cover is decorated with a border of white ovals, outlined in black. The corners are banded in black and white as shown in Fig. 7.

Care and Storage of Camp Equipment

A slovenly sportsman misses much of the joy of the man who takes pride in giving his outfit the proper care, not only during its period of use, but also during the winter, when occasional overhauling serves to keep one in touch with sports of other seasons. And a very real joy it is, each article recalling an experience as one examines it minutely for a possible rust spot, scratch, or injury.

Tents usually come in for much abuse, which shortens their life considerably. Cotton duck molds quickly, and rots if left rolled up damp. Care should be taken, therefore, to insure its perfect dryness before storing. Silk and silk-composition tents, being thoroughly waterproof, are almost as dry after a rain or dew as before, so may be packed for moving at any time. But all tents and tarpaulins should be washed and dried carefully after the season's use.

Blankets absorb much moisture, and should be shaken and spread out over bushes to dry in the sun, at least once a week. In the cold nights of late summer, the increased warmth of blankets after drying is considerable.

Pack straps and ropes should not be left exposed to the weather. They speedily become hard or brittle; squirrels like the salt they can obtain by chewing the leather, and if left on the ground in a rabbit country, the straps are soon cut into bits. Hang the leather goods in the peak of the tent, keep them away from fire, and oil them occasionally.

A canoe should not be left in the water overnight, or at any time when it is not in use. Simply because use makes it wet, a canoe should not be left so any more than a gun should be left dirty, or an ax dull. If on a cruise with a heavy load, pile the stuff on shore at the night camp, and turn the canoe over it. If a canoe is permitted to remain in the water unnecessarily, or its inside exposed to rain, it soon becomes water-soaked and heavy for portage, besides drying out when exposed to the sun, and developing leaks.

Small punctures in the bottom of a canoe may be mended with spruce, tamarack, or pine gum, melted into place with a glowing firebrand, held close, while blowing at the spot to be repaired. Torn rags of canvas-covered bottoms may be glued with the softer gum of new "blazes," gathered with a knife or flat stick.

While traveling on shallow streams, the bottom of a board canoe develops a "fur" of rubbed-up shreds. Every night these should be cut short with a sharp-pointed knife, to prevent a shred from pulling out and developing into a large splinter. The paddles, and the setting pole, unless shod with iron, become burred at the ends and require trimming down to solid wood

The track line, if in use, is wet most of the time, and unless dried frequently, becomes rotten. Every tracker knows the grave danger with a rotten line in a rapids.

During the winter the canoe should be scraped and sandpapered, bulges nailed down, permanent repairs made to the covering, and the canoe painted on the exterior and varnished on the interior.

The average fisherman is an enthusiast who needs no urging in the matter of caring for his outfit, and the user of firearms should profit by this example. Even if not a shot has been fired from a gun all day, moisture from the hands, or from the dampness in the woods, or marshes, may cause rust spots, or corrode the bore. Rub an oily rag through the bore and over the outside of the gun every evening, before laying it aside.

Cleaning rods are safer and more thorough in cleaning the bore than the common mouse string, which may break when drawing a heavy piece of cloth through, causing much difficulty. A wooden rod, preferably of hickory, is best, although the metal rod is stronger for use in small bores, but care must be taken not to wear the muzzle unduly. The hunting weapons should be carefully overhauled before storing them, and given a coat of oil to protect the metal parts from rust.— A. M. Parker, Edmonton, Can.

Useful Periscope Which a Boy Can Make

Mention of periscopes is quite common in the reports from European battle fields; such a device in a simple form can be made easily by boys who have fair skill with tools. The illustration shows a periscope which may be used for play, and has other prac-

tical uses as well. In a store or other place where a person on duty cannot watch all parts of the establishment,

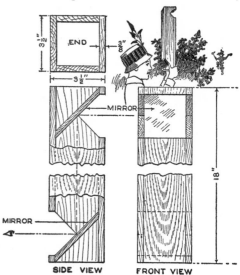

SIDE VIEW   FRONT VIEW
This Simple Periscope Is Useful Both for Play and Practical Purposes

such a device is convenient in that it will reflect persons entering the door. As a toy or for experimental purposes the periscope shown has many possibilities, and will appeal to youngsters.

It consists of a square box, 18 in. long, open at the ends. It is 3½ in. wide and made of wood, ⅜ in. thick. A mirror is fitted at an angle of 45° near one end of the box or tube, as shown in the sketch. The front of the mirror is opposite a three-cornered opening in the box which extends across one side. The opposite end of the tube is also fitted with a mirror in the same manner, except that the front of the mirror faces to the opposite side of the box at which there is also an opening. In using this device, the user sights from the point indicated by the eye. The image is reflected in the mirror at the top and thrown onto the lower mirror, where it may be seen without exposing the head above the level of the lower opening. It is this application of reflection by mirrors that makes it possible for soldiers to see distant objects without exposing themselves to fire, by the use of the periscope.

Trap for Coyotes

Two coyotes, resisting capture in a hole under a lava ledge, were hauled forth quickly when the device shown

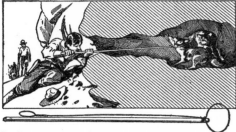

By Drawing on the Wire the Coyotes were Brought under Control and Forced from Their Lair

in the sketch was used, after other means had proved ineffective.

I made a snare of baling wire and attached it to a pole, 6 ft. long, running the wire down from the loop to the end of the handle. The loop was made about twice the size of the coyote's head, and, by drawing on the wire at the handle, the animals were placed under control and held at a safe distance.—Contributed by Milton Barth, Geyserville, Cal.

Utilizing an Empty Paste Pot

The type of paste pot that contains a central well to hold the paste brush is not always cast aside when empty. A pincushion built to fit the outer ring,

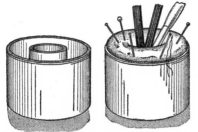

A Pincushion Built to Fit the Empty Paste Compartment Makes a Useful Article of the Paste Pot

or paste-holding section, is added and the brush holder is used to hold pens and pencils.

⊄A magnet may be used to advantage in picking up tacks or small nails which have been scattered on a floor.

A Colonial Mirror Frame

Black walnut, or mahogany, is the most effective wood to use in making this simple but artistic frame. It requires a very small amount of stock, and what is used should be of a good quality and carefully worked to the given dimensions with keen tools. The stock required for the frame is as follows:

Black walnut, or mahogany:
2 pieces, 27½ in. long, 1⅜ in. wide, and ¾ in. thick.
1 piece, 22 in. long, 1⅜ in. wide, and ¾ in. thick.
1 piece, 9¼ in. long, 1⅜ in. wide, and ¼ in. thick.
White holly:
1 piece, 27½ in. long, 1½ in. wide, and 1/16 in. thick.
Picture board:
1 piece, 25 in. long, 9 in. wide, and ⅛ in. thick.

The dimensions for the walnut or mahogany pieces are rough sizes, oversize to allow for the planing to the dimensions given in the sketch. The white holly may be procured smoothly planed on both sides and of the exact thickness required. The picture backing may be purchased in almost any store that sells frames. It is usually rough pine and inexpensive.

The first operation is to plane the frame pieces on one side and edge, using great care to insure both being perfectly straight and the edge square with the face. Gauge for, and plane to the thickness required, although this need not be exactly ⅝ in. as called for, but if the stock will stand 11/16 in. or ¾ in. do not take the time to cut it down to ⅝ in. The little cross rail must be exactly ⅛ in. thick, as it is to be let ⅛ in. into the rabbet cut for the glass, which makes it come ⅛ in. back from the face of the frame when it is in place. Plane all of these pieces to the width, 1⅛ inch.

For cutting the rabbet, a plow, or a ¾-in. grooving, plane is the best tool to use, but if neither is available a rabbet plane can be used. Be sure to plane the rabbet square and to the lines gauged for the depth and width. To groove the pieces for the holly strips a special tool is required. This may be made of a piece of soft sheet steel or iron, which must be of a thickness to correspond to that of the holly. A piece 2½ in. long, and of almost any width, will answer the purpose. File one edge of the metal straight, and cut saw teeth in it by filing straight across with a small saw file. Remove the burr raised by the filing by rubbing each side on an oilstone. Drill two holes in it for fastening with screws to a piece of hard wood. The wood serves as a fence, and if properly fastened to the metal, the teeth should cut a groove ⅟₁₆ in. deep and ³⁄₁₆ in. from the edge. The holly strip should fit the groove tightly so that it can be driven home with light taps of a hammer. It is well to try the tool on a bit of waste wood first to see if it cuts the groove properly.

The holly is cut into strips, ⅛ in. wide, with a slitting gauge. An ordinary marking gauge, with the spur filed flat on each side to make a sharp, deep line, will do very well for this work. The gauging is done from both sides of the piece to make the spur cut halfway through from each side. Before the slitting is attempted, one edge of the piece is first straightened. This is readily accomplished with a fore plane, laid on its side and used as a shoot plane. The strip to be planed is laid flat on a piece of ⅞-in. stock with one edge projecting slightly. This raises it above the bench and allows the fore plane to be worked against the projecting edge.

The strips should be applied to the groove to test the fit, and if found to be tight, they must be tapered slightly by filing or scraping the sides. If the fit is good, hot glue may be run into the grooves with a sharp stick, and the strips driven into place. They will project above the surface slightly, but no attempt should be made to plane them off flush until the glue has become thoroughly hardened; then use a sharp plane, and finish with a scraper and No. 00 sandpaper.

The miters are cut in a miter box, or planed to the exact 45° angle on a miter shoot board. Before gluing the corners, the recesses are cut for the

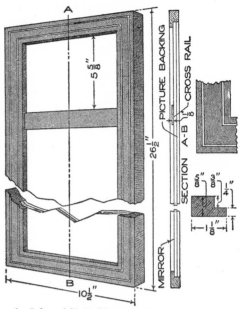

An Inlay of Holly Makes an Exceedingly Pretty Frame of Colonial Design for a Mirror

cross rail, but it must not be put in place until the corners of the frame have been fastened and the glue given time to dry.

The frame may be given either a dull or bright finish. The dull finish gives a rich appearance and is very easy to apply. Give the completed frame one coat of white shellac, and when it is dry, rub the surface with very fine sandpaper until it has a smooth finish. Finish with any of the prepared waxes, being careful to follow the directions furnished.

Before putting the board back of the mirror, be sure to place two or three sheets of clean paper on the silvered surface. The picture board is fastened with glazier's points, or with small bung-head wire nails. The back is finished by gluing a sheet of heavy wrapping paper to the edges of the frame. If the wrapping paper is moistened with a damp cloth before it is applied, it will dry out smooth and tightly drawn over the back.

⁋An emergency penholder may be made by fixing the pen on the end of a pencil with a clip, or small rubber band.

A Turntable Stand for Potted Flowers

Potted flowers, if kept in the house, tend to grow toward the light. From time to time the pot should be turned. To do this more readily, the turntable stand shown in the sketch was designed. It is made up of a low, four-legged taboret upon which a 12-in. disk of 1-in. wood is fixed with a screw. A thin wooden washer, sandpapered and shellacked, insures easy turning.

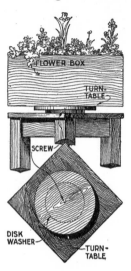

Rectangular boxes or circular jars look equally well upon the stand, the beauty of which depends much upon its workmanship and finish.—Edward R. Smith, Walla Walla, Wash.

Index for Popular Mechanics Magazine on Bookshelf

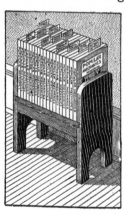

As it would be impossible to keep in one's mind all the good ideas given in Popular Mechanics Magazine, I preserve my back numbers for reference, and, to reduce the amount of index and page searching, I have adopted a convenient plan of indexing any special article to which I expect to refer. I write, on one end of a strip of paper, the page and date of issue of the article. I place these slips at the proper pages in the magazine, so as to project at the top, as shown, making reference easy. If a slip drops out, it is quickly replaced. When articles are no longer needed for reference, the slips are removed. By keeping the magazines on an open shelf they can be reached handily, for reference as well as for removing them from the shelf.—J. E. McCormack, Haliburton, Ont., Can.

Trimming Board with Foot Control and Counterweight

A trimming board with the knife operated by a pedal, leaving both hands free to handle the work, is a device appreciated by photographers, commercial artists, and others who have more or less heavy paper or cardboard to trim. An ordinary trimming board is mounted on a packing box of suitable

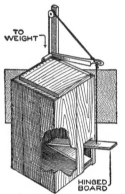

dimensions, as shown. A pedal is hinged to the bottom of the box, and a heavy cord, or wire, runs to the handle of the knife, so that pressure on the pedal operates the knife. The latter is returned by a weighted rope, passing over a pulley attached to a standard at the rear corner.—H. F. Blanchard, New York, N. Y.

Folding Card Table Handy for Invalid in Bed

When it is necessary to serve a meal to a patient in bed, and no invalid table is at hand, use an ordinary folding card or serving table. Unfold one end over the bed, and a splendid substitute table is provided, useful for reading, games, and other purposes. If the table rests too low on the bed, adjust it with cushions. This is far more convenient than using a tray.—John P. Rupp, Norwalk, Ohio.

A Pigeon House

By Robert Baker

PIGEON houses need not be eye-sores, as is often the case, but may be made to harmonize with the surroundings, adding beauty to a dull spot, and even making the grounds of a home more attractive. The house described will accommodate 20 pigeons, and additional stories of the same type may be added to provide for more. Nearly all of the wood necessary may be obtained from boxes, and the other materials are also readily available at small cost. The construction is such that a boy handy with ordinary carpentry tools may undertake it successfully.

The house is constructed in general on principles used in buildings, having a framed gable roof, rough-boarded and shingled. The interior arrangement is original, being based on the Indian swastika or good-luck sign. While the construction is simple, it must be carried out systematically. The process outlined also follows in general the typical methods in building construction.

The foundation need not be considered, since the house rests upon a post, and the construction thus begins with the lower story. The floor and the ceiling are similar in construction, as shown in Fig. 1. In framing them into the lower story, as may be observed in Fig. 8, the cleats are placed below on the floor and above in the ceiling. The construction is identical, however. The cleats are fastened to the boards with screws, although nails, clinched carefully, may be used. The 4-in. hole at the center should be made accurately, so as to fit the shoulder portion at the top of the post, shown in Fig. 2. The latter may be cut of a length to suit; about 9 ft. will be

found convenient. The notches in the top of the post are to fit the ridge pole and center rafters of the roof frame, as shown in Fig. 10. They should not be made until the house is ready for the roof boards.

The pieces for the compartments, as arranged on the floor in Fig. 3, are made next. Figs. 4 and 5 show the detailed sizes of these pieces, of which four each must be made. The sizes shown must be followed exactly, as they are designed to give the proper space for entrances and to fit around the 4-in. square hole, through which the post is to fit. The pieces marked A, B, and C, in Figs. 4 and 5, correspond to those similarly marked in Fig. 3.

The pieces are nailed together to form the swastika in the following manner:

Mark the pieces A, B, and C, as shown. Measure 4 in. from one end of each piece marked A, and square a pencil line across, 4 in. from the end. Arrange the pieces in pairs. Place one end of one piece against the side of the other piece in the pair, so that the pencil line is even with the end, permitting the 4-in. portion to project. Nail both pairs in this position. Then fit the two parts together to form a 4-in. square in the center, as shown in Fig. 3.

Fit the pieces C to the pieces B at an angle, as shown in Fig. 3, trimming off the projecting corners where the pieces are joined. Nail them together, and they are ready to be fixed to the end of the pieces A, already nailed. By nailing the joined pieces B and C to the end of the pieces A, as shown in Fig. 3, the swastika is completed. Fix it into place, with the center hole exactly over the square hole in the floor, by means of nails or screws driven through the floor.

Two small strips must now be nailed to the floor at each side of the swastika. They should be exactly 4½ in. long, and are to hold the slides, Fig. 9, which shut off the various compartments. The slides are shown hanging by chains in the headpiece of this article, and are shown in place in Fig. 8.

Fix the ceiling into place in the same manner, being careful that the square holes fit together, and that the cleats are on the upper side. Turn the construction over and fix into place the small strips for the slides, as was done on the floor.

The fixed screens, Fig. 6, and the doors, Fig. 7, are constructed similarly. They are built up of ½-in. wood, and vary in size to fit their respective places in the framework. Observe that the fixed screens are ¼ in. higher than the doors, and that they are fastened between the ceiling and floor, bracing them. The wire grating is ½-in. square mesh, and is fixed between the pieces of the doors and the screens when they are built up.

The doors are shown secured by combination strap hinges, bent over the baseboard. Plain butts may be used and the lower portion of the hinge covered by the baseboard, a recess being cut to receive the part covered. In the latter instance the doors should be fixed into place immediately after the screens are set. Catches and chains may then be placed on the doors. Next nail the baseboards into place. They are 2½ in. wide and may be mitered at the corners, or fitted together in a square, or butt, joint. The latter joint may be nailed more readily.

The slides, shown in Fig. 9, may now be made and fitted into their grooves. The handles are made of strips of band iron, drilled for screws and bent into the proper shape. It is important that the slides be constructed of three pieces, as shown, so that they will not warp or curve from exposure. The main piece is cut 7¾ in. long, and the strips, ½ in. square, are nailed on the ends.

The construction of the framing for the roof should next be taken up. This probably requires more careful work than any other part of the pigeon house, yet it is simple, as shown in Fig. 10. Note that the rafters are set upon a frame, or plate as it is called, built up of pieces 3 in. wide. It should be made ¼ in. wider and longer on the inside than the ceiling board, so as to fit snugly over it. The joints at the corners are "halved" and nailed both ways. This gives a stronger structure than butting them squarely and nailing them. The end rafters should be fitted in before fixing the others. It is best to make a diagram of the end of the roof framing on a sheet of paper, or a board, and to fit the rafter joints in this way before cutting them. The rafters are then nailed into place.

The "rough boards" to cover the rafters may now be nailed down. They are spaced ½ in. apart so as to permit thorough drying, as is done in larger buildings. They project 2 in. beyond the ends of the plate frame, supporting the rafters. A ½-in. strip is nailed over the ends to give a neat finish. The roof may be shingled, or covered with tar paper, or any roofing material.

Nail a 1-in. strip under each end of the roof and nail the gable ends into place. One gable end is provided with a door, as shown, and the other has an opening fitted with a wire screen of the same size as the door.

The gable story rests on the lower story, and the notches in the top of the post should fit snugly to the ridge and center rafters, as shown in Fig. 10. This will aid in supporting the house firmly. If additional stories are added it would be well to place a post at each corner of the house. The upper story may be removed for cleaning, or for transporting the house.

FIG.1

POST
6" SQUARE
FIG.2

FIG.3

FIG.4

FIG.5

FIXED SCREEN
FIG.6

DOOR
FIG.7

SLIDE
FIXED
SCREEN
FIG.8

RIDGE

POST

RIDGE

FIG.10

The House Accommodates 20 Pigeons; Additional Stories may be Added. Fig. 1, Floor and Ceiling; Fig. 3, Arrangement of Compartments; Fig. 8, Lower Story Assembled Ready for Roof Story; Fig. 10, Framing of the Roof; Fig. 11, Side View, Showing Spacing of Roof Boards; Fig. 12, End View, Showing Trim and Door on Gable End

FIG.9
SLIDE

FIG.11

1"SQ. STRIP

FIG.12

131

The post should be sunk into the ground about 2½ ft. and set into a concrete foundation, if convenient. This will insure a more nearly permanent as well as a more rigid support. Care should be taken that the post is set plumb and this can be accomplished if a plumb bob is used. The post should be braced to keep it vertical, particularly if a concrete foundation is poured and tamped around it.

The construction should be painted two coats, inside and out, of a color to harmonize with buildings or other surroundings.

The cost of building the house shown in the illustration was $3.50, and by using tar paper for the roof, and discarded wire mesh, hinges, and other fittings, this may be reduced considerably.

⁋In matching molding into corners it is often difficult to make miter joints. The molding may then be "coped" together by matching the end of the piece to be joined to it with the curves or surface of the molding. A coping saw is used in sawing the irregular end.

Roses Tinged Blue by Chemicals

Roses may be colored without any detrimental effect by placing their stems in a solution of 100 cubic centimeters of water, 2 grams of saltpeter, and 2 grams of an aniline dye. A centerpiece of roses colored to represent the national colors was made in this way and proved very effective as a table decoration. A convenient way to color the flowers is to place their stems in a test tube containing the mixture.—Contributed by Chester Keene, Hoboken, N. J.

Making Photographic Trays

Serviceable trays for use in developing and printing photographs may be made quickly of cardboard boxes of suitable sizes. Where one is unable to transport readily a full photographic equipment these trays will prove convenient as well as inexpensive. They are made as follows:

Procure boxes of proper sizes and see that they have no holes or openings at the corners. Melt paraffin and pour it into the box, permitting it to cover both outside and inside surfaces If the paraffin hardens too rapidly the box may be heated and the work completed.—Contributed by Paul A. Baumeister, Flushing, N. Y.

Camp Lantern Made of a Tin Can

Campers, and others who have need of an emergency lantern, may be interested in the contrivance shown in the sketch, which was used in preference to other lanterns and made quickly when no light was at hand. It consists of an ordinary tin can, in the side of which a candle has been fixed. A ring of holes was punched through the metal around the candle and wires were placed at the opposite side for a support. The glistening interior of the can reflects the light admirably.—Contributed by F. H. Sweet, Waynesboro, Va.

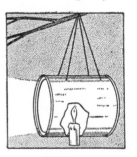

⁋Interior woodwork may be made proof against ordinary flame by coating it with silicate of soda, known as water glass.

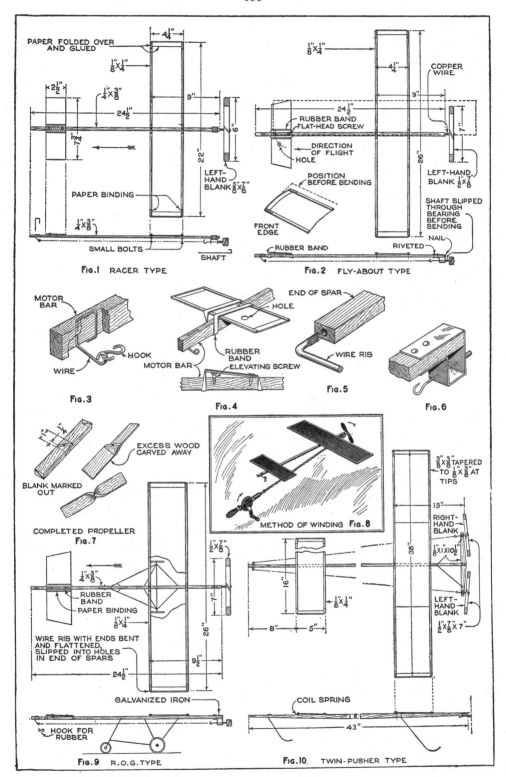

PAPER FOLDED OVER AND GLUED

$\frac{1}{8}" \times \frac{1}{4}"$

$\frac{1}{4}" \times \frac{3}{8}"$

PAPER BINDING

$\frac{1}{4}" \times \frac{3}{8}"$

SMALL BOLTS

SHAFT

FIG. 1 RACER TYPE

$\frac{1}{8}" \times \frac{1}{4}"$

COPPER WIRE

RUBBER BAND
FLAT-HEAD SCREW

DIRECTION OF FLIGHT

HOLE

POSITION BEFORE BENDING

FRONT EDGE

RUBBER BAND

RIVETED

NAIL

LEFT-HAND BLANK $\frac{1}{2}" \times \frac{7}{8}"$

SHAFT SLIPPED THROUGH BEARING BEFORE BENDING

LEFT-HAND BLANK $\frac{1}{2}" \times \frac{7}{8}"$

FIG. 2 FLY-ABOUT TYPE

MOTOR BAR

HOOK

WIRE

FIG. 3

END OF SPAR

HOLE

RUBBER BAND

MOTOR BAR

ELEVATING SCREW

FIG. 4

WIRE RIB

FIG. 5

FIG. 6

EXCESS WOOD CARVED AWAY

BLANK MARKED OUT

COMPLETED PROPELLER
FIG. 7

METHOD OF WINDING FIG. 8

RUBBER BAND
PAPER BINDING

$\frac{1}{8}" \times \frac{1}{4}"$

WIRE RIB WITH ENDS BENT AND FLATTENED, SLIPPED INTO HOLES IN END OF SPARS

$\frac{1}{4}" \times \frac{3}{8}"$

GALVANIZED IRON

HOOK FOR RUBBER

FIG. 9 R.O.G. TYPE

$\frac{1}{8}" \times \frac{1}{4}"$

COIL SPRING

$\frac{3}{8}" \times \frac{3}{8}"$ TAPERED TO $\frac{1}{8}" \times \frac{3}{8}"$ AT TIPS

RIGHT-HAND BLANK

$\frac{1}{8}" \times 1" \times 10"$

LEFT-HAND BLANK

$\frac{1}{2}" \times \frac{7}{8}" \times 7"$

FIG. 10 TWIN-PUSHER TYPE

The Four Styles of Model Airplanes Illustrated have Proved Satisfactory in Flight, and Possess Features That Make Them Easy to Build. The Line of Flight and Altitude of the Models are Regulated by Adjusting the Wings and Elevators. The Driving Power is Furnished by Rubber Bands

HOW TO BUILD
MODEL AIRPLANES

By DONALD W. CLARK

THE model airplanes illustrated, while they do not exemplify the very best performance or design, nevertheless have proved to be very satisfactory in flight, and have structural features that make them easy to build. Because of this simplicity of design, they will appeal to the person who likes to build, whether he has had experience in this line or not, and there is little doubt of his ability to complete the style of his choice, provided he has the necessary patience and consideration for detail.

Since the "racer," Fig. 1, the "fly-about," Fig. 2, and the "rise-off-ground,"

and the middle of the spars, wrapping two or three thicknesses of paper around them; this is to prevent splitting. From a piece of soft wire, ⅟₁₆ in. in diameter, cut a piece 1 in. longer than the distance between the centers of the spars, and bend right angles ½ in. from each end. Flatten the ends of the wires on an anvil or vise, lay the spars flat on a smooth surface, and insert the short ends of the wires into the holes in the ends of the strips, forcing them in as far as they will go, as indicated in Fig. 5. When these wires are bent to control the direction of flight, they will stay so, because they are

Above, Left: The "Fly-About" Model Shown with Two Extra Wing Ribs; Center: The "Twin-Pusher" Type with a Cruising Radius of from 600 to 800 Feet. Below, Left: This View of the Fly-About Model Shows How the Motor is Assembled; Center: A Tiny Model Compared with a 12-Inch Rule; It is Driven by a Two-Inch Propeller. Right: One of the Model Planes in Flight

Fig. 9, types are nearly alike, they will be described first. The wings of all three are built up in the same manner, using materials of the same kind, but differing in their dimensions. To build the wings for either model, two strips of white pine, basswood, or spruce are selected, a trifle over the required length, planed down to measure exactly ⅛ by ¼ in., and then cut to length. Mark the middle of the strips, and drill a ⅟₁₆-in. hole through each. In the center of each end of the strips drill a ⅟₁₆-in. hole ½ in. deep. Next, cut several strips of tough paper 1 in. wide, coat them with glue, and bind each end

soft and because the flattened ends prevent them from moving up and down. This method bends the rear spar a little, but it has proved satisfactory on many models. After the wing frame has been assembled, lay it on a piece of tough paper, mark the outline, and then cut it out, leaving a ½-in. margin on all sides. Now coat the underside of the frame with glue, place it on the paper, having the margin even all around, and work out the wrinkles, being careful not to bow in the spars. Cut the ends of the paper to fit between the spars, coat them with glue, fold over the wires, and stick to the top side of the paper. After the glue has dried thoroughly, lay the wing on a smooth board and trim off the surplus paper. Punch a hole through the paper binding over the holes in the spars, and give the whole a coat of waterproof varnish to make an exceedingly tough and durable unit.

For the motor bar, cut a piece of pine or spruce, ¼ by ⅜ by 24½ in. Lay it flat, drill a ¹⁄₁₆-in. hole about ¾ in. from one end, and another ½ in. from the first. Bend a piece of soft wire as in Fig. 3, slip it through the holes, and bend the ends as shown. The propeller bearing on the racer is made from a piece of sheet metal, drilled for the ¹⁄₁₆-in. diameter shaft, and fastened to the bar as shown in Fig. 6, although it can be bound to the shaft with glued paper; the latter will, perhaps, be the better way for the beginner. Figure 7 shows how to make the propeller. The shaft should be made of wire, bent as shown in Fig. 1, and slipped through the hole drilled in the center of the propeller; by indenting the hub a little with the short end of the shaft, the exact position for the extra hole can be found. When this is drilled, slip the shaft through the hub again, pressing the short end into the extra hole; this prevents any chance of the propeller turning on its shaft. Another way is to flatten the end of the shaft and force the widened part into the wood, parallel with the grain.

The elevator, Fig. 4, is made of pine or basswood, bound in the middle with glued paper, as described for the wing spars and indicated by the shaded section in Fig. 1. Plane down the wood to ¹⁄₁₆ in. in thickness, and cut a piece of the proper size; for the racer, it should measure 2½ by 8 in., for the fly-about 2½ by 8½ in., and for the rise-off-ground model 3 by 9 in. Bind the edges with a strip of paper and varnish. The elevator of the racer is not movable, but is attached with two round-head screws to the motor bar. This makes it impossible to alter the angle of the elevator on this model without removing it from the bar. Two or three small washers underneath the forward edge serve to place it at the correct angle. On the fly-about and rise-off-ground models, however, the elevator is adjustable, and is fastened with but one screw near the rear edge, which serves as a pivot, and one rubber band; the latter holds the elevator down against a flat-headed screw located under the front edge, and also keeps it straight. To increase or decrease the angle of the eleva-

tor, merely turn it to one side, so that the hole shown in the drawing will come over the screw head. Then, with a small screwdriver, turn the screw in or out as needed, and allow the elevator to return to its normal position. The rubber band also prevents breaking of the elevator by absorbing some of the landing shock.

The rise-off-ground model, Fig. 9, is just like the fly-about, except that it has a little more surface than the latter, and the wing is set at an angle large enough to give it a good lift. This is done with small washers, or coiled wire of sufficient thickness to raise the front wing spar about ⅛ in. The drawing shows how the landing gear, which consists of three wheels mounted on hard wire, is attached. Make the wheels of cigar-box wood, drill the centers, and use the same size of wire for the axles as for the wire supports, the ends of which should be looped around the axles. The small front wheel must be a little lower than the others, so that the forward end of the motor bar will be higher than the rear when the model is resting on the ground. The diameter of the front wheel is 1½ in., and the larger ones are 2 in. in diameter.

The "twin pusher," Fig. 10, is more elaborate than the other three, but is not beyond the amateur's ability. Its elevator does not swing, but it is made in exactly the same manner as the other models. This is also true of the wing, except that in this case the spars are ⅜ in. square in the middle, tapered down to ⅛ by ⅜ in. at their tips. Two extra wire ribs are placed 6 in. from the center, between the spars. The holes are drilled and the ribs inserted before the paper wing covering is applied. The motor bar is of pine or spruce, ½ by 1 by 43 in., shaped as shown in the drawing. Make the front hook from a 4-in. length of wire, insert it in a small hole drilled near the front of the bar, and bend the loops. The crossbar that takes the bearings should be of ash or other hardwood, and be braced with ¹⁄₁₆-in. hard wire. The bearings are strips of sheet metal, bent to U-shape and riveted to the crossbar, and hard wire is used for the landing skids. The front edge of the wing should be raised about ¹⁄₁₆ in. The elevator is adjustable, and is attached to the motor bar by two screws; the front one runs through a small coil spring between the spar and motor bar, and furnishes a means of changing the angle of the elevator.

The power on all these models is furnished by rubber bands, about ³⁄₆₄ in.

thick, ⁹⁄₁₆ in. wide, and 4 in. long, linked together chain fashion, so that three of the bands will only make a length of 6 in. instead of 12 in. This method allows broken bands to be replaced with new ones quickly and easily.

A wire hook, inserted in the chuck of a hand drill, as shown in Fig. 8, will serve as a winder. After linking the bands together on the model, release the front end of the "motor," and hook it to the drill. Stretch the rubber to about twice its length and turn until about half wound. Then keep turning, but gradually release the tension so that the rubber will be straight when fully wound, and hook on again. The number of turns needed will be found through experience; the twin model will stand more than 1,000 turns to each propeller, which means that with a

gear ratio of 4 to 1 on the drill, the handle will have to be turned 250 times. The two propellers on this model must revolve in opposite directions.

To launch, hold the motor bar with the right hand, just ahead of the wing, and the propeller with the left. Then, with a quick, upward push, send the airplane into the air. If it has a tendency to climb too steeply, the elevator should be lowered a little, and if it loses altitude, the elevator should be raised enough to correct this fault. With both wing tips flat, the planes will have a tendency to turn to the right. Curving the right-wing tip down a little will give a straightaway flight; a left turn can be made by curving the right tip down still more. Several trial flights will probably be necessary before the proper adjustment is obtained.

Simple Hanger for Cut Films

An ordinary coil spring, such as used for holding window curtains, can be made into a holder for developed exposures

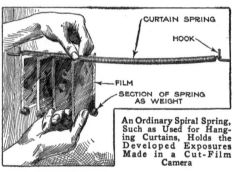

An Ordinary Spiral Spring, Such as Used for Hanging Curtains, Holds the Developed Exposures Made in a Cut-Film Camera

from a film pack. The spring is suspended between two hooks, as shown. The films are inserted between the coils, the compression of the spring preventing them from slipping out. To prevent the cut films from curling, ½-in. sections of a larger and heavier spring may be slipped over the bottom to serve as weights.— W. W. Baumeister, Cambridge, Mass.

Registers Used as Ventilators

Adjustable steel hot-air registers can be used as a means of ventilating the garrets of houses having poor ventilation under the roof. They are very easily put up and are neat in appearance. An opening is cut in the gable end of the house and the register plates fitted and screwed to the siding.—Louis M. Steffen, Dayton, Ohio.

Testing for Leaky Tire Valves

Automobile tires, suspected of having a slow leak at the valve, can be tested without removing the tire from the rim. A convenient method for detecting such leaks consists in filling a small bottle with water and immersing the valve stem in it. The wheel having the suspected valve is turned so that the valve is at the top, and the valve stem is inserted into the neck of the bottle. Leaks will manifest themselves by a series of bubbles.

Neat Tips on Wires

In order to have the wiring look neat on a radio set, or other instrument being made, all wires with visible connection to binding posts should be wound with small bare wire to prevent the insulation from raveling out. Fig. 1 shows the insulation removed from the wire, the winding of the tip started

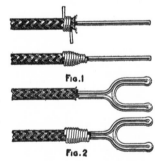

with a piece of bare wire of small diameter, and the neat finished effect. Fig. 2 illustrates the method of making a tip for connecting to another style of binding post. The latter is a connection that is especially useful when the wire is to be frequently connected to or disconnected from a binding post.

A Neat Radio Interrupter

The radio interrupter illustrated is made from common materials and pre-

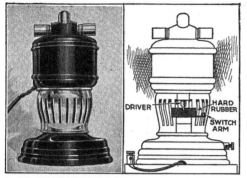

An Electrically Operated Interrupter for the Amateur Radio-Transmitting Set That is Made from an Electric Motor, the Metal Guard of an Old Casserole, and Miscellaneous Electrical Parts

sents a very attractive appearance when finished.

The base, made from a wood or hard-rubber fixture base, has mounted in its exact center a discarded switch arm, knob, and bearing, from which a wire connection runs to a binding post at the edge of the base. Soldered to the end of the switch arm, which clears the slotted metal guard by 1/8 in., is a small piece of spring brass, about 1/40 in. thick by 1/4 in. wide. This is so arranged as to make and break contact with the small silverplated casserole guard, which is mounted between the motor and base, and to which a binding post is soldered for the second connection. The shaft of the small fan motor used is connected at the lower end to the knob of the rotating switch arm by means of a tube, forced over the shaft and flattened at the end. This end fits into a slot in the hard-rubber switch knob, so that there is no direct electrical connection between the switch and motor.

In case a device such as this is to be used on a high-voltage lighting or other circuit, it will be best to insulate the motor from the slotted casserole cover. This can be done by placing heavy-paper washers and bushings on the motor bolts.
—R. U. Clark III, Newton, Mass.

Keeping Water Out of Automobile Casings

Automobile owners are frequently troubled by water in the tire casings, admitted through the opening for the valve stem. This water and the rust that results, cause much wear on the inner tube.

An excellent method of preventing this trouble is to cut several washers, about 2 in. in diameter, from an old inner tube. A small hole is cut in the center of each, so that it can just be forced over the valve stem, making a tight fit. When the rim is put in place on the wheel, the washers will be compressed in the space between the felly and rim, making it impossible for the water to enter.

An Improvised Dark Room

For changing plates, loading holders, and even developing negatives, when there is no dark room at hand, an ordinary coat and a wash boiler, box, or similar light-tight vessel, can be made to answer. The coat is buttoned up and the skirts are securely tied around the top of the wash boiler, in the case shown in the photo, or the box, after the plates and holders and a small light-tight box have been put in. The operator runs his arms

down through the sleeves of the coat and changes the plate or removes exposed films from the film pack, placing them in the box. In the case of film packs, single films can be developed by placing them in a tumbler of developer that is mixed as for a tank; other tumblers are provided for clear water and for the fixing bath.—Hallie H. Holt, Salem, Ill.

Feed Box That will Not Overrun

It is frequently desired to have a feed or salt box in a feed lot or pasture. Some animals seem to delight in upsetting the usual type of box, but one made according to the drawing, with the sides extending beyond the corners, will be proof against the mischievous antics of the stock.

The pieces for the box are cut just twice the length required for the actual dimensions of the box.

Homemade Greeting Cards

BY W. C. HARRIS

IF one wishes to get away from hackneyed printed or engraved greeting cards during the holiday season, it is a comparatively simple matter to make most attractive cards by photography.

The first step, if elaborate cards are to be made, is to build up the "dummy." The negative, or a finished print, is laid on a sheet of drawing paper and a layout of the lettering, etc

is made around it. The lettering, monograms, etc., that are to appear on the design may be drawn on separate pieces, and the whole arranged to best advantage around the photo. When the desired arrangement is completed, the loose parts are inked in, stuck down on the bristol board together with the print, and a negative made of the whole layout. This is used to print the greeting

Left: Greeting Card with White Letters, Made In on Enlargement. Right: Letters Printed in Black by Double Printing. Below: A "Dummy," from Which Large-Sized Cards are Made

Combining Two Films. Top: Letters Inked

cards. If the lettering is to appear on the print itself, an enlargement, of reasonable size, is made on a good smooth-finish paper, upon which the lettering is inked, making the sentiment expressed suit the picture used, as shown in the illustrations. From this, a copy negative, the exact size of the desired card, is taken, and contact prints made. These should be mounted on rough cover paper, contrasting in color with the print.

The lettering may be printed in white through some thin part of the original negative, by making a positive of the lettering, on commercial film, and then laying it over the picture negative so that the lettering comes in the proper position. It is necessary to make the positive film larger than the negative, so that no edge may show in the finished print, and to keep it all clear except for the letters, by masking the negative from which it is printed. The letter positive and picture negative are bound together, and the cards printed through both.

It is best to make the card prints on double-weight, linen-finish buff or white paper, and it is well to make the prints to suit a particular size of envelope, as it is usually easier to do this than to find an envelope to suit the print.

Should any difficulty be encountered in making the prints stay flat, it may be overcome by dampening the backs, and putting them between blotters, under a weight, and allowing them to dry.

Ventilating Doors for Garage

The garage doors shown in the illustration are somewhat unusual in that a lattice gate is hinged to the edge of one

A Lattice Gate, Hinged to One Garage Door and Hooked to the Other, Keeps Undesirables Away When the Owner is Working on the Car

door, and, when the doors are opened, may be hooked to the edge of the other.

This gate is made of wood strips, set in a light frame at an angle of 45°. When the doors are shut, or opened full, the gate is folded back against the door to which it is hinged. This arrangement admits plenty of light and air while working on the car, and at the same time prevents the entrance of unwelcome visitors.

Renewing a Radio Crystal

When the galena crystal of a small radio set has become dull and dirty, so that it is almost impossible to find a sensitive spot, a simple but very effective method can be used to renew it.

The mounted crystal should be held over a flame, in an old spoon, until the metal begins to run. The crystal can then be turned over with a piece of wire, thus exposing an entirely new sensitive area. A mold can be used to keep the melted metal in the original shape and size, so that it will fit in the detector cup. Care should be taken not to heat the metal too much beyond the melting point, as this will impair the sensitiveness of the crystal.

Improving the Hand Drill Press

The photograph shows a method by which a small hand drill press can be more conveniently operated. With the drill press as bought, a clamp is provided for holding the work, as it is necessary to use both hands, one to turn the crank and the other to feed the drill by pressure on the lever. By drilling a hole in the feed lever and attaching a strap, or chain, running from the lever to a pedal on the floor, the necessary pressure of the drill against the work is obtained by pressing 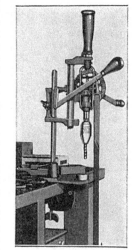 on the pedal. This makes it unnecessary to clamp the work, except in special cases, and leaves one hand free to hold it while the other is used to turn the drill.—C. R. Gains, Colfax, Ia.

A HOMEMADE MOTION-PICTURE CAMERA

By R.E.Best

THE making of a moving-picture camera and projecting apparatus that will do work quite favorably comparable with that of high-priced machines, is not so difficult an undertaking as might at first appear, nor does it involve great expense. By the use of ordinary tools, the expenditure of a little time, and by careful work, a very satisfactory piece of apparatus may be produced.

The camera here described uses standard motion-picture film, and will contain 100 ft. at one loading. The magazine can be removed and the camera itself used to project the pictures, which will be found to be practically as satisfactory as the use of a separate projector. The dimensions are taken from a camera made by the author;

The Finished Camera Set Up and in Operation: Note the Heavy Tripod, Which Is Necessary to Insure Steady, Even Pictures

a slight variation may in some cases be necessary, depending on the material used, and the focal length of the lens.

Standard motion-picture film is 1⅜ in. wide, and perforated on both edges. The pictures are 1 by ¾ in., and there are 16 pictures to the foot. The film must be moved past the lens with an intermittent motion at the rate of a foot a second. A rotary shutter is provided to cut off the light when the film is moving, in order to avoid blurring the pictures. There are several ways in which this intermittent motion may be produced, but

the one described will be found the most practical for amateur construction.

The lens can be borrowed from an ordinary camera. Since the pictures are of such small size, the lens must be of short focal length, preferably from 2½ to 4 in. A combination of two lenses of 6 or 7-in. focal length will give the desired result. The box is best made of maple about ¼ in. thick, although any thin close-grained boards will do. The joints should be mortised, as shown in the illustrations, to prevent any possibility of light leakage. The gears can be obtained for a modest sum from any house selling standard gears. The sprockets can be made by one who has access to a lathe, but it is better to buy them from a motion-picture supply house. The other materials can be obtained from any hardware dealer. The following materials will be needed:

1 6-in. gear, 32-pitch.
2 3-in. gears, 32-pitch.
1 ¾-in. gear, 32-pitch.
1 pair bevel gears, ½ by 2 in.; that is, ratio 1 to 4.
2 standard motion-picture sprockets, 16-tooth, with idlers.
1 ¼-in. gear, 48-pitch.
1 rack, 6 in. long, ⅛ in. square, 48-pitch.
1 bushing, 5/16-in. inner, and ⅜-in. outer diameter.
3 lb. babbitt.
Brass rod, 5/16 in., ¼ in., 3/16 in., and ⅛ in. in diameter.

First make the box, 12 by 7 in. inside dimensions, the front portion containing the mechanism to be 7¾ in. deep and the rear portion, or magazine, 5 in. deep. One side of the front portion should extend 2 in. above the top of the box to form one side of the finder. A groove, ⅛ in. deep and ½ in. wide, should be cut in the top and bottom of the front, 2¼ in. from the right side, looking from the back, before the box is assembled, to allow the plate that holds the gears, sprockets, etc., to slip in and out. A rectangular hole, 3½ by 4⅛ in., is cut in the front to allow the focusing device to be set in, and an opening for the inspection door in the left side. All joints should be glued and made light-tight. The rear portion should have a ¾-in. strip set around the inside to project about ¼ in., so that when the two parts of the box are brought together, this strip will come inside the front and back portions, and make the joint light-tight. It would be well to face the edge of the box with felt.

Set the lens in the front of a box, constructed of ¼-in. material, 3 by 3⅝ in. by 1½ in. in outside dimensions. The bottom edge of this box should extend about ⅛ in. beyond the sides. A piece of the 48-pitch rack, ⅛ in. square and 1½ in. long, should be set flush in the center of the bottom. This box should slide quite easily in a 3½ by 3½ by 4⅛-in. box which is set in the front of the camera. The base of this outside box is bored in the center to take the pinion, and a hole drilled at right angles to this for the pinion shaft; grooves are also cut in the sides to fit the projecting bottom of the box holding the lens. The box should have a block, 2¾ by 2¼ by 1¼ in., bored with a hole somewhat larger than the outside diameter of the lens, fastened to the inside of the back. This block should be fitted so that the box containing the lens will not bind in sliding back and forth. The back of the outside box should have a hole cut in it, of sufficient size to let the light from the lens fall on the film in the film gate, yet small enough to keep the light from spreading over other parts of the camera.

The pinion can be made by soldering a ¼-in., 48-pitch gear to one end of a short piece of ⅛-in. brass rod. If a knurled brass button for the other end cannot be obtained, a button can easily be cast from babbitt.

This arrangement will be found to permit the lens to move quite freely in focusing and will admit no light to the camera except through the lens. When the camera is complete, the proper position of the lens for objects at various distances can be determined by trial, and the positions marked.

The gears, sprockets, film gate, striker, shutter, etc., are all supported by one piece or plate which can be removed from the box as a unit. Make this plate of ½-in. hardwood, 12¼ by 6 in. Fit it to slide in the grooves in the camera box, where it can be fastened in place with screws when the mechanism is finished. Another plate of the same material, 11¾ by 6 in., should be fastened to the first by means of blocks, separating the two plates 1⅝ in. The blocks must be located carefully, to avoid interference with the gears, and held in place by screws. Mark out the location of the four shafts, and bore holes in both plates, ½ in. or larger in diameter. Make bearings of babbitt or heavy brass, bored to the size of the shaft and shaped so they can be fastened to the wooden plates by means of screws. These bearings can be moved until the gears mesh properly before they are permanently fastened in place.

The shaft to which the crank is attached should be of ⁵⁄₁₆-in. brass rod. This shaft carries the 6-in. gear and the 2-in. bevel gear, and extends outside the camera to take the crank.

The crank should be turned at the rate of two turns a second, and, as the sprockets have 16 teeth, representing 3 in. of film, or four pictures to the turn, the sprocket shafts should turn twice as fast as the crankshaft in order to feed the film at the rate of 16 pictures a second. These two shafts are ¼-in. brass rod and carry 3-in. gears. The lower shaft should have a 2-in. pulley placed between the sprocket and the bearing plate to drive a round belt running to the take-up reel. This pulley can be easily made of two circular pieces of cigar-box wood, glued together and grooved.

A ³⁄₁₆-in. hole should be bored in the edge of the bearing plate for the shutter shaft, and the plate must be cut away in the center to allow the ½-in. bevel gear attached to this shaft to mesh with the 2-in. bevel gear on the crankshaft. The hubs of all the gears should be fitted with setscrews, and it is well to file flats on the shafts under these screws to prevent slipping.

The idlers should be attached to strips of heavy sheet brass and placed against the sprockets in the position shown. The strips should be pivoted at one end on a screw, and held in place with small coil springs.

The shutter should be cut from a 4-in. disk of sheet brass, bored with a ⅛-in. hole in the center and soldered to the end of the shaft, which should be shouldered to fit the hole. It must be designed in such a manner that it will shut off the small pieces to the back. It is quite important that the wheel be balanced to prevent vibration, which produces unsteady pictures.

The film must be held stationary and perfectly flat when the exposure is made.

Upper Left: Inside View c Camera. Upper Right: Camer and Cover, Showing Camer Reel Plate. Lower Left: Camer Set Up. Lower Right: Projec tion Assembly

light while the film is ii motion. If the blade each cover 60° of th circle they will be founa to be sufficiently large. The shutter has two blades and consequently should turn four times as fast as the crankshaft.

The striker serves not only to pull the film down through the film gate as it is taken up by the lower sprocket, but acts also as a flywheel. For this reason it should be heavy and well-balanced. It should be 2¾ in. in diameter, by ⅜ in. thick, and may be cast from babbitt in a wooden mold (do not use pine or any wood containing pitch for molds). The arm that comes in contact with the film may be made of sheet brass, and soldered to the edge of the wheel. It should be filed away slightly in the middle for a distance of 1 in. so that it will only come in contact with the edges of the film and not scratch the pictures. After the striker has been bored and fitted to the edge of its ³⁄₁₆-in. shaft, it should be trued up on a lathe if one is available, and carefully balanced by boring holes or soldering

It must also be allowed to move quite easily when pulled down by the striker. This is done by means of the film-gate assembly, which consists of three parts: the film gate, cast in babbitt, the pressure plate, and the cover, made of heavy sheet brass. The film gate is fastened rigidly to the main bearing plate by means of screws; the pressure plate holds the film against it and is allowed to give slightly as the film moves through. One edge of the cover fits into a groove or notch in the film gate, and the other edge is held down by means of a thumbscrew.

It requires a little ingenuity to build the mold to cast the film gate, but by combining blocks of wood of the proper size and shape, a mold can be built up which will make a very nice casting. When the casting is finished, smooth it up with a file and cut a groove in the base with a hacksaw for the edge of the cover. The surface around the "frame" and along the edges of the passageway for the film must be carefully smoothed, as these parts come in contact with the film, which is very easily scratched. The rectangular hole, or frame, should be 1 in. by a little less than ¾ in., to avoid any overlapping of pictures.

The pressure plate should fit quite loosely in the film passageway in the film gate. Its upper edge is curved to avoid scratching the film, and it is fitted with four short posts of ⅛-in. brass rod, which reach through the cover and come in contact with flat springs. These springs keep the pressure plate tightly against the film, yet allow it to give when uneven parts of the film, such as splices, come through. The frame for the picture is cut through both the cover and pressure plate as well as the film gate, to allow the light to go through in projecting the pictures. The side of the pressure plate which comes in contact with the film should be highly polished to avoid any possibility of scratching the film.

The cover consists of a piece of sheet brass to which are attached the springs that hold the pressure plate against the film. These springs are pieces of watch spring, supported in the center and inclosed at each end. The center supports can be cut from sheet brass and soldered in place. The little boxes inclosing the ends of the springs can also be made of sheet brass, and soldered to the cover. This arrangement will be found to hold the springs in their proper place without the necessity of boring the hard steel. The springs must be bent until they put the proper pressure on the film; this will have to be determined by experiment.

Build a box, 2 by 2½ by 8 in. inside dimensions, on top of the main camera box, to inclose the finder. Fasten a small double-convex lens of 5 or 6-in. focal length in the front of this finder, and leave a hole, about 1 in. in diameter, behind the lens to admit light. A lens suitable for this purpose can be obtained from any optician or may be taken from an old bicycle lamp or flashlight. The screen for the finder can be made by removing the emulsion from a piece of film and roughening it with fine sandpaper. It should be placed in the finder at the focal distance of the lens. When the camera is all complete and assembled, a black border should be painted on the screen to allow only those parts of the image to show that are thrown on the film in the frame by the main lens. The back of the finder should be inclosed to shut out the light, and a peephole, about

1 in. across, should be cut in it. A brass hood will add to the appearance and allow the finder to be closed when not in use.

The camera is now ready to assemble. Fasten the film gate to the bearing plate in such a position that the frame will come directly behind the lens, and at such a distance that the range of the focusing device will permit objects at any distance to be focused on the film. Put in all the shafts, gears, sprockets, etc., attach the crank, and adjust the bearings until all the gears mesh properly and run smoothly.

Now remove the crank and crankshaft, insert the bearing plate in the camera box, and fasten it in place by screws at top and bottom. Locate the position of the crankshaft on the side of the camera, and bore a hole somewhat larger than the shaft. Cast a circular piece of babbitt, 1 in. in diameter and ³⁄₁₆ in. thick, and bore a ⁵⁄₁₆-in. hole in the center. This is placed on the outside of the box, the shaft inserted, and the gears set in place. The circular piece of babbitt can then be fastened to the outside of the camera by means of screws, to act as a bearing and prevent the leakage of light around the shaft. The shutter should be properly timed when the bevel gears are meshed.

A 6 by 11 by ³⁄₁₆-in. plate should be fastened to the back of the main bearing plate for the purpose of holding the supply and take-up reels. Thumbscrews should be used for this purpose, as the plate must be easily removable when the camera is used for projection purposes. The pulley on the take-up reel must, of course, line up with that on the shaft in the camera, and a small belt be fitted.

The fastenings that hold the cover should be designed to pull the two parts of the camera tightly together. Fasteners can be made of heavy wire and small pieces of strap iron, as shown in the illustration. The tripod must be of heavy construction and quite rigid, as a wobbly tripod will produce jerky, unsteady pictures.

By removing the plate holding the supply and take-up reels, the camera can be used for the projection of the finished film, and if the work has been carefully done, the resulting picture will be as clear and steady as those made by professional apparatus. The best light for home use is a 250-watt stereopticon lamp, set behind a pair of condenser lenses. These can be taken from a stereopticon, or purchased at the optician's. The box must be well ventilated, and should be

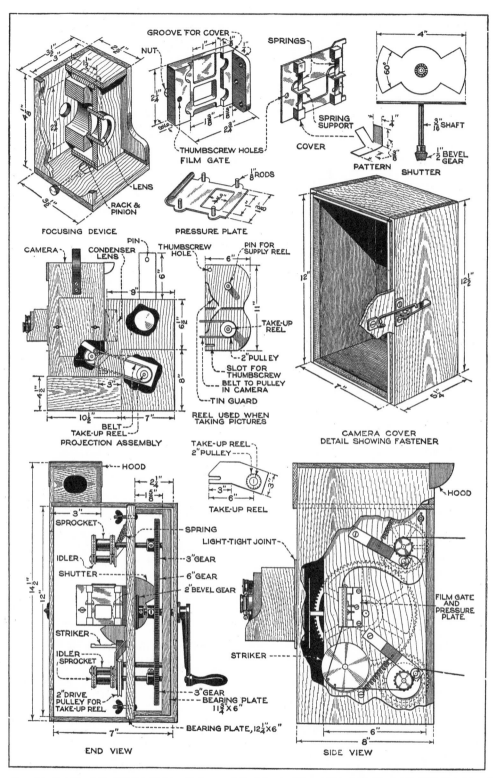

GROOVE FOR COVER

NUT

THUMBSCREW HOLES
FILM GATE

SPRINGS

SPRING
SUPPORT

COVER

PATTERN

SHUTTER

60°

4"

3/16" SHAFT

1/2" BEVEL
GEAR

LENS

RACK &
PINION

FOCUSING DEVICE

1/8" RODS

PRESSURE PLATE

CAMERA

CONDENSER
LENS

PIN

THUMBSCREW
HOLE

PIN FOR
SUPPLY REEL

TAKE-UP
REEL

2" PULLEY

SLOT FOR
THUMBSCREW

BELT TO PULLEY
IN CAMERA

TIN GUARD

REEL USED WHEN
TAKING PICTURES

BELT
TAKE-UP REEL

PROJECTION ASSEMBLY

12

12 1/2"

CAMERA COVER
DETAIL SHOWING FASTENER

HOOD

TAKE-UP REEL
2" PULLEY

TAKE-UP REEL

HOOD

SPROCKET

IDLER

SHUTTER

STRIKER

IDLER
SPROCKET

2" DRIVE
PULLEY FOR
TAKE-UP REEL

SPRING

LIGHT-TIGHT JOINT

3" GEAR

6" GEAR

2" BEVEL GEAR

STRIKER

3" GEAR
BEARING PLATE
11 1/4" X 6"

BEARING PLATE, 12 1/4" X 6"

FILM GATE
AND
PRESSURE
PLATE

END VIEW

SIDE VIEW

Complete Details of the Motion-Picture Camera and Projector: Upper Drawings Show the Construction of
the Focusing Device, Film Gate, and Shutter; Center Drawings, Projection Apparatus, Picture-Reel Plate,
and Camera Cover, and Lower Drawings, Side and End Views of the Camera Completely Assembled

144

lined with asbestos, as it gets very hot if used for any length of time. A supply reel and a take-up reel must be arranged, and the latter can be run by a belt from the pulley in the camera.

If one expects to develop and finish the film in his own dark room, some sort of apparatus must be constructed to handle the wet film. In handling small quantities, of 50 ft. or less, the most economical way is to construct a cylindrical reel 15 to 24 in. in diameter and about 2 ft. long. Let this rest in a frame so that, as it turns, it dips into a pan containing the chemical. The film can be wound on the reel, emulsion side out, and passed through the successive baths of water, developer, and hypo, and then dried without being disturbed.

When the negative film is developed and dried, it must be printed on positive film before being projected. It can be printed in the camera by running the positive and negative films through together past a light, but it will be found more economical to send the negative to a laboratory, where the printing will be satisfactorily done for a few cents a foot.

A Practical Sliding Gate

The sliding gate shown in the drawing has decided advantages over many styles

A Type of Sliding Gate Which Is Neat in Appearance and Easy to Handle, as There Is Very Little Friction in the Parts

of similar gates now in use. It is neat in appearance and easy to open and close, as there is very little friction.

If a clear 10-ft. opening is desired, the posts on each side of the driveway should be about 11 ft. apart. One gatepost is made double to prevent the gate from being dislodged, and has a pulley in the center, to carry the top bar of the gate. Both upper bars are 22 ft. long and extend to another post, placed 11 ft. from the double one, where they are held in place by two grooved pulleys, as shown. A crosspiece must also be provided on the gate to prevent it from sagging.

An Improvised Developing Tray

A photographer, having broken one of his glass developing trays, and being unable to obtain another immediately, made a tray of paraffin to serve as a temporary substitute.

After melting about a pound of paraffin, it was poured into a large pan of hot water, until the layer of paraffin floating on the surface was about ¼ in. thick. When the paraffin had cooled down until it was fairly hard but still pliable, it was removed and pressed into a glass developing tray, the inside of which had been moistened with water to prevent the paraffin from adhering to it. The surface of the layer of paraffin was smoothed, and the edge cut even with the glass edge. After hardening, the paraffin was removed and found to form a serviceable tray that could be cleaned very easily and was not affected by the chemicals.

Tool for Removing Headlamp Rims

The removal of the bayonet-lock type of headlamp rim is one of the stubborn jobs that tries the patience and skill of practically every auto repairman or service man. The drawing shows a tool by means of which the rim is gripped with a uniform pressure at all points, and it is only necessary to grip and turn the handles to loosen even a tightly "frozen" rim, without damage.

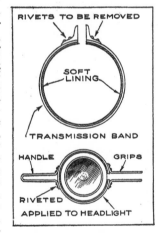

The tool is made from the transmission band of a light automobile, and fitted with a new, soft, white lining. The ears of the band are knocked off and their place taken by a pair of flat-iron grips, and on the opposite side of the band a handle is riveted.

A SIMPLE BATTERY RADIOPHONE

By F. L. BRITTIN

A N experimental radiophone transmitter, operated on dry cells and a storage battery, can be built for approximately $35, if all the parts are purchased; this does not include the cost of the batteries, and can be reduced somewhat by using homemade parts. With an instrument of this type, conversation can be carried on over distances up to 10 miles. This is the type of radiophone the amateur should build while learning

parts of this set, such as the inductance, the fixed and variable condensers, the transmitting tube, choke coil, modulation transformer, microphone, and the resistances.

The panel, which is made of ¼-in. bakelite, is 10 by 12 in. in size, and is attached to the ¾ by 8 by 10-in. wooden base by means of angle brackets. The binding posts for the batteries, aerial, and ground are arranged on the front of

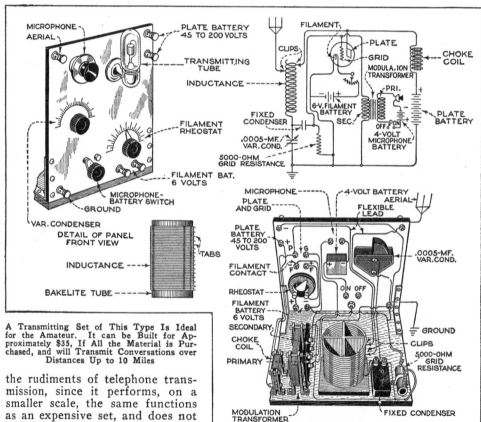

A Transmitting Set of This Type Is Ideal for the Amateur. It can be Built for Approximately $35, If All the Material is Purchased, and will Transmit Conversations over Distances Up to 10 Miles

the rudiments of telephone transmission, since it performs, on a smaller scale, the same functions as an expensive set, and does not make learning so costly, when accidents occur to the instrument.

Before attempting to construct this set, the reader should study the wiring diagram carefully, and should, at least in a general way, understand the principles involved in radiophone work, besides knowing the functions of the various

the bakelite panel, as shown in the drawing. The inductance is wound on a threaded bakelite tube, 4½ in. in diameter and 7 in. long. If not already threaded when purchased, the tube should be put on the lathe and threaded, so that it will take

146

80 turns of stranded "litzendraht" wire, or the same amount of No. 16 bare copper wire, if the former is not available. At every five turns, a small tab of sheet copper is soldered to the wire, so that the clips of the aerial and plate can be snapped on at different points, to vary the inductance.

The lower end of the inductance coil is connected to one terminal of the variable condenser and also to the fixed condenser. The other terminal of the variable condenser is connected to the ground post. The second terminal of the fixed condenser is connected to three points: to the grid, to one terminal of the secondary coil of the modulation transformer, and to the grid resistance. The remaining terminals of the secondary coil of the transformer and the grid resistance are connected to the ground.

The six-volt storage battery is connected to the filament, care being taken to connect the negative terminal directly to the filament binding post, and the positive terminal to the filament rheostat. The negative terminal must also be connected to the ground.

The B-battery consists of at least four 22½-volt units, such as used for receiving sets. Several units can be used, as the greater the voltage supplied to the plate, the greater the distance spanned by the message. For this reason the plates on transmitting tubes are designed to stand high voltages, as high as 500 volts for the 5-watt size used in this set. The units are connected in series, the negative terminal being connected to the ground line, while the positive lead passes through the choke coil to the plate, and also to the inductance coil, to which it is attached by means of a helix clip.

The simple matter of connecting the microphone circuit remains. This is done by placing a 4-volt battery, consisting of 3 dry cells, an on-and-off switch, the microphone, and the primary coil of the transformer in series on a separate circuit, as shown.

The variable condenser used is of .0005-mf. capacity, panel-mounting type, and should be able to stand a high voltage. A stop should be used to prevent it from shorting at 90°, as some condensers on the market are designed to do.

The fixed condenser is also of .0005-mf. capacity, especially designed for high voltage. The type used consists of several strips of copper foil between mica, the whole being bolted between two metal plates.

The 5,000-ohm grid resistance is of standard vitrified type. The filament rheostat is of 7 ohms' resistance. The choke coil is of the 150-milliampere type. If the builder desires, he may use an anticapacity switch in the B-battery circuit, to permit throwing the battery current from the transmitting to the receiving set. The latter, of course, is a separate instrument.

It should also be kept in mind while wiring, that, although the wires are shown running parallel in the diagrams, for the sake of simplicity, they should be kept at right angles to each other as much as possible, in order to reduce induction.

LIST OF MATERIALS

1 ¼ by 10 by 12-in. bakelite panel.
1 ¾ by 8 by 10-in. wood base.
2 6 by 6-in. angle brackets.
6 binding posts.
1 filament rheostat.
4 contact points.
2 switch levers, with knobs.
1 .0005-mf. variable condenser.
1 .0005-mf. fixed condenser, C. W.-
  type, for high voltage.
1 microphone.

1 5-watt transmitting tube.
1 4½ by 7-in. bakelite tube.
1 5,000-ohm vitrified resistance unit.
1 modulation transformer.
1 6-volt storage battery.
3 dry cells.
4 22½-volt B-batteries.
1 150-milliampere standard choke coil.
3 helix clips.
Litzendraht wire for inductance, or No.
  16 copper wire.

1 tube socket, for panel mounting.

Useful Hints for the Motorist

A blow-out patch in an automobile casing will last much longer if a piece of old inner tube is cut and laid between the patch and the inside of the casing. This will prevent the ragged edge of the hole in the casing from wearing through the patch.

The solderless connections that hold the flexible copper tubing to the gasoline tank and carburetor, often have a tendency to work loose, owing to the constant vibration. By substituting a piece of graphited asbestos-cord packing, such as is used to pack water-pump glands, for the slip ring, it will be found that the joint will not leak, nor will it work loose as quickly as before.—Wm. Byers, Kansas City, Mo.

Simple Method of Making Candles

Very serviceable candles, that produce a flame four or five times as large as that of an ordinary wick candle, can be made in the following manner: About 1 lb. of paraffin—such as is used for canning purposes—is melted in a flat pan. A long sheet of tissue paper, about 6 in. wide, is grasped at each end, and is slowly passed through the melted wax several times, each layer being allowed to cool before the paper is passed through again. When the thickness of the wax is about ⅛ in., the sheet is allowed to cool, and is then cut into pieces from 6 to 12 in. long, depending upon the size of candle desired, and rolled into cylinders.

When dipping, the wax should not be too hot, as it would then melt the layers previously applied, instead of adding a new one. It is best to melt the wax over water, at a temperature of about 212° F. To expose the paper wick for lighting, revolve the end of the candle in a flame until some of the paper lights; it is then ready for use.

A Simple Picture-Frame Clamp

The illustration shows a device that has proved to be very satisfactory for clamping glued picture frames. Four pieces of hardwood, 3 by 1¼ in., are cut and grooved as shown. Two pieces of soft steel, or iron, about 5 in. long and 1 in. wide, are drilled at each end. A bolt, about 8 in. long, threaded full length, passes through the center of one plate,

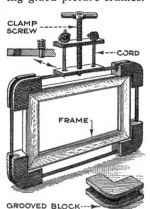

which is drilled and tapped to fit it. The bolt is then turned down at the end and seated in a hole drilled through the other plate. A rope is passed around the grooves and through the holes and knotted securely, and the apparatus is assembled as shown. The turning of the bolt will tighten the rope and bring equal pressure upon all four corners of the picture frame. If the clamp screw and the rope are long enough, various sizes of frames can be clamped in the device.

"Moving-Picture" Toy for Children

A very interesting "moving-picture" toy for the small child can be made of cigar

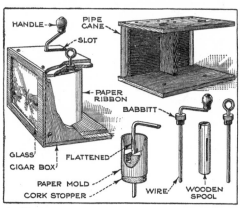

A "Moving-Picture" Toy for the Small Child, That Is Instructive as Well as Amusing, and That can be Very Easily Made from Scrap Materials

boxes, some wire, babbitt or lead, and a few pieces of pipe cane.

A rectangular opening is cut in the bottom of a cigar box, of the size made to contain 100 cigars, and a piece of window glass, cut to fit, is placed behind the opening and held in place by tacks. A frame, shown at the upper right, made of cigar-box wood, fits neatly into the box; this holds the picture ribbon against the glass and carries the spools on which the ribbon is wound. A piece of pipe cane is mounted at each corner of the frame, to serve as a roller.

The spindles, one of which has a crank formed on one end, are made of No. 9 galvanized wire, and are provided with babbitt or lead "keys" to turn the wooden spools on which the ribbon is wound. To make these keys, a portion of the wire should be flattened, as shown. A cork stopper is then taken, and, after a groove is cut in the top and a hole drilled through the center, it is pushed on the wire directly underneath the flattened portion. By wrapping heavy paper around this cork, a cup is formed, into which the babbitt can be poured. The wooden spools upon which the ribbon is wound can be made from old film spools, cut at one end to fit the babbitt key.

Children can be amused for hours with this little toy, which can be made instructive as well as amusing. Pictures cut from the comic or rotogravure sections of newspapers, and pasted to the ribbon in order, make very interesting moving pictures of this kind, although, of course, any suitable pictures may be used.

A Substantial Electric Radiant Heater

Many electric radiant heaters are made with porcelain cores in several parts, and these are very easily broken. After breaking one of these cores several times, it was found possible to remodel the heater so that the risk of breakage was reduced to a minimum.

A hard-maple block, 5 by 2½ in., was turned in a lathe to the shape shown, the middle part being made 5 by 1½ in., and the ends turned down to about 1⅞-in. diameter. On each side of this block a hole was bored. A porcelain wall receptacle was obtained, and all the copper lining, except the lower part of one pole, was removed, a type of receptacle that had the screw hole in the center being used. A common porcelain

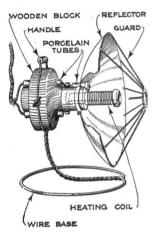

WOODEN BLOCK — REFLECTOR
HANDLE — GUARD
PORCELAIN TUBES
HEATING COIL
WIRE BASE

tube was wrapped on one end with asbestos, and this end was glued firmly into the wall receptacle. The wire of the old heating coil was wound around the tube, care being taken to have sufficient space between the turns to prevent short circuits. Both ends of this wire were brought back to the pole screws of the receptacle, one end through a small porcelain tube in the reflector, and the other through the center of the heating-unit tube. The outer covering of the extension cord, but not the insulation, was cut away, and the single conductors were brought through porcelain tubes fixed in the holes in the wooden block, and connected to the pole screws of the receptacle. Soldering the wires to the pole screws made them secure. A thin collar and a clamp screw were used to attach the reflector to the receptacle. An attractive and rigid base was then made of No. 4 copper wire, twisted as shown, and attached to the wooden block with thumbscrews, in such a manner that the heater could be adjusted and locked at various angles. The wire was also brought around the top to hold a wooden handle. The exposed pole screws should be taped and varnished to protect the user.

Book Ends Made of Angle Iron

A neat pair of book ends can be made from two pieces of angle iron, the size varying according to the size of the volumes to be supported. The pieces are ground smooth on a grindstone or emery wheel, and then painted or enameled to give them a neat appearance. A layer of felt, glued to the bottom, will prevent them from scratching the finished surface of a table or desk.

Dead-End Switch for Inductances

The efficiency of a radio-receiving set can often be improved to a considerable extent by the use of a dead-end switch to short-circuit the unused turns of the inductance. The switch illustrated can be added to the set without disturbing the existing arrangement, and at a negligible cost. It consists simply of a piece of spring brass, bent as indicated, and soldered to the knob shaft, at the back of the panel. Small pieces of brass, bent to a right angle, are drilled and fastened under the head of each contact-point screw, to provide positive contacts for the spring-brass wiper. A little care

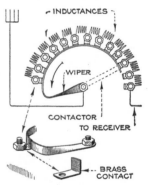

INDUCTANCES
WIPER
CONTACTOR
TO RECEIVER
BRASS CONTACT

is necessary to insure that each contact piece is placed correctly, and that the wiper is in the proper position, relative to the regular switch contactor, to short-circuit the unused turns and to touch all the contact pieces.

This switch can be used with any multipoint switch in which the points are arranged in an arc, and where contact is made by a rotating switch arm.

Waterproofing Felt Hats

An old hunter uses the following method of waterproofing felt hats: A solution made by dissolving half a bar of common soap in water is liberally applied to the inside of the hat. When this is dry, a second solution, made by dissolving a cupful of powdered alum in water, is brushed on over the first. The soap is thus hardened, and becomes waterproof.

Motor Wheel Serves as Home-Shop Power Plant

By L. B. ROBBINS

A POWER PLANT that will operate small drills, grinding wheels, lathes, and similar machines, in the home workshop, can be readily made from one of the popular motor wheels, in the manner shown in the illustrations. When the bicycle is laid up for the winter, the wheel can be taken off and used in the shop.

First, choose the location best suited for the installation; this should be an overhead beam or joist running at right angles to the bench. Set the machine to be operated on a bench, so that it will not interfere with other work, and then locate the motor wheel back of the center line of the machine, if possible, so that the belt will run at an angle. The motor wheel is suspended from the ceiling by a flat-iron hanger; this should be long enough so that, when it is fastened to the beam and the motor attached, there will be at least 15 in. between the top of the tank and the ceiling. This arrangement provides room for filling the tank with fuel. One of the hanger legs is a straight piece of iron, while the other is offset so that the bottom ends are 12 in. apart. Drill three holes in the top of the legs and one ½-in. hole an inch from the bottom of each; the latter is for the pivot bolt. Attach the legs to opposite sides of the overhead beam or joist with bolts or lagscrews, and make sure that the pivot holes are exactly in line with each other. Then drill a smaller hole in the straight leg, 15 or 16 in. above the

pivot hole, and bolt a diagonal iron brace to it and the beam—it may be found necessary to use two such braces, and one between the legs, as shown, to keep them rigid and perpendicular.

Place the end of the motor wheel that attaches to the bicycle between the legs, and pivot it to them with a bolt made

A Motor Wheel, Detached from the Bicycle When the Summer Is Over, and Installed above the Bench, Provides a Small Power Plant for Driving the Tools in the Home Workshop

from a piece of ½-in. round iron rod, threaded at both ends and provided with nuts and washers. Set a piece of pipe over the bolt to serve as a spacer; this should be long enough to keep the motor wheel bearing against the straight leg.

Drill two or three small holes in the rear end of the mudguard and bolt on the adjusting lever shown in the drawing. This lever is made from a piece of heavy iron, twisted at one end and curved to fit the mudguard. Holes should be drilled in the iron to correspond with those drilled in the guard, in addition to a hole in the end for the pull wire.

Next, a guide is made from a piece of round rod, the ends of this being flattened, and bolted to the beam directly over the adjusting lever. With the motor wheel blocked horizontally, adjust the guide so that the bottom will be about 3 or 4 in. below the adjusting lever. Forge the sides of the guide together, so that they just allow the lever to slide easily between them without undue play.

and under the metal catch shown, the motor will be raised about 3 in. above the bottom of the guide.

Lastly, make a back frame of flat iron. This should be bolted to the bent leg of the hanger at the pivot hole, and to the end of the motor wheel below the adjusting lever. This is done by removing the bolt that binds the engine crankcase to the motor guard, slipping the iron against them, and bolting all three together with a longer bolt. Then bolt a properly bent piece of flat iron onto the middle of this back frame, at right angles to it, and bolt or rivet the end to the mudguard underneath the fuel tank; this forms a rigid frame and keeps the motor wheel vertical

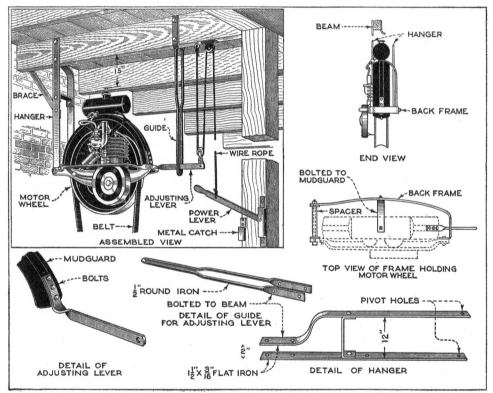

Insert: The Complete Assembly. Upper Right: End and Plan Views of Back Frame and Brace. Lower Left: Detail of Adjusting Lever. Center: Construction of Guide. Lower Right: Assembly of Hanger

Attach a pulley and a screweye to the overhead beam, and a pulley to the end of the adjusting lever. Then fasten a length of small wire rope to the screweye and reeve it through the pulleys as shown. Lead the end of this rope down to a pivoted lever that is fastened to a convenient post, or to the wall, and attach it so that, when the lever is up, the motor wheel will be lowered to the bottom of the guide; when the lever is down

and always in line with the machine below.

Place the control lever in a convenient position and remove the tire from the rim, to leave a pulley large enough for high-speed work. Belt the machine so that the belt is loose when the lever is raised; this allows the motor to run without operating the machine. As soon as the power lever is pulled down and caught under the catch, the belt will be tightened and the machine will turn. In this way the engine can be kept running while the machine is stopped for any reason.

The engine is started by setting the controls and spinning the rim, and the speed can be regulated to suit the work in hand by means of the spark and throttle controls.

Starting a Saw Cut Smoothly

Everyone using a handsaw has, at one time or another, had trouble in getting the cut started in the wood, as the teeth catch in the edge and either stop the saw or mar the edge. A remedy for this is to file smaller teeth on the last two inches of the blade nearest the handle. This may be done by filing the point off the teeth so that a notch can be cut in each of the old teeth, thus making two teeth from each one. A few strokes with these small teeth will start the cut easily. If more convenient, the small teeth can be filed at the outer end of the saw.— Chas. N. Shaw, Jr., Oxford, Ga.

Joining Rag Strips for Weaving

Much of the time and labor required to sew the strips of rags together, which are to be woven into carpet or braided into rugs, can be saved by a simple and easily made device. A small piece of thin, flat steel is bent at right angles, the upper end pointed and sharpened, and a slot or eye cut in it, through which the strips may be inserted. Two holes are also drilled into the horizontal part for attaching to a

table or wooden block. In use, the strip to be joined is laid over the end of the other and both pushed down over the sharpened point; then the loose end of the former piece is brought through the eye, and pulled up close. When this has been done, the junction of the two strips is pulled up from below, the portion in the eye passing through the slit. After the strip has been pulled out of the eye and drawn up tightly, the resulting joint will be almost perfectly flat.

Auxiliary Piping Prevents Overheating

A means of preventing an automobile engine from overheating can be provided by adding an auxiliary outlet pipe in the

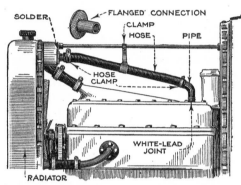

Preventing Overheating of Automobile Engine by Providing an Auxiliary Pipe: Thus the Water from the Rear Cylinders Passes Directly into the Radiator

manner suggested in the drawing. A hole is drilled and tapped near the rear of the cylinder-head casting to take a 1 or 1¼-in. pipe; this is bent as shown, and, after coating the threads with white lead, screwed into the hole. A hole is cut into the back of the upper header on the radiator, and a flanged connection of the type shown is soldered, or, better, brazed to it. The two connections are united by a section of rubber hose, which, if inclined to vibrate too much, can be supported from the radiator stay rod. The extra pipe allows the hot water from the rear cylinders of the engine to pass to the radiator without passing over the forward cylinders.

Repair for Broken Eyeglass Spring

Wearers of pince-nez or "nose glasses" often find that one or both of the small springs on the bridge of the nosepiece are broken, or have become so weak that the glasses slip off. Of course, the proper thing to do is to take the glasses to a jeweler or optician for repair, but, in an emergency, resort can be had to the method illustrated. One end of a small rubber band is slipped over one of the guards and wound around the

bridge often enough to obtain the proper tension, when the opposite end is pulled over the guard on the opposite side.— Truman R. Hart, Ashtabula, Ohio.

Match-Box Holder and Lighter

Easily made from a piece of sheet metal, the match-box holder and cigar lighter

A Match-Box Holder and Cigar Lighter That Prevents the Small Boxes of Safety Matches from becoming Crushed in the Pocket, and Provides a Windshield That Makes It Possible to Get a "Light" Even in a High Wind

illustrated makes it possible to carry the small boxes of safety matches in the pocket without crushing, and provides a windshield for the flame so that it is easy to get a "light" even in a strong wind. The sheet-metal strip from which the device is formed is cut to the width of the match box, which is about 2¼ in. long. A slot is cut in the middle of the pattern, through which the match to be struck is inserted. Also, an ear, about ⅞ in. long, is provided on what is to be the bottom. After the sheet-metal form has been cut, it is bent as shown, and the ear is turned up and soldered to the sides. The holder illustrated is embellished with a foreign coin.

A Two-Handled Cord Reel

A reel that will save much time in the winding of kite strings, fishlines, clothes-lines, and other cords, can easily be made from a piece of board of the proper length. The ends are cut out to hold the cord and to prevent it from slipping off, and handles are fastened on opposite edges of the board, near each end, as illustrated. After attaching the cord to the reel it may be wound up easily and rapidly.

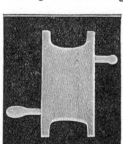

Pulling Light Car Out of Mud

An owner of a light automobile has used a novel idea for extricating his car from bad mudholes. He raises the engine hood and passes one end of an ordinary clothesline through the footboard, tying it to the clutch pedal. The engine is then speeded up, the driver walks to the loose end of the rope, throws it over his shoulder, and pulls. This throws in the clutch and applies power to the rear wheels, which, together with the force of the pull and the lessened weight in the car, suffices to pull the car out in most cases. As soon as the pull on the rope is relaxed, the clutch will throw itself out, and the car stops. The emergency-brake lever should be kept in the neutral position during the operation; this allows the low-speed band to engage, but prevents the pedal from flying back into high and stalling the engine.

Keeping Insects from Light Bowls

A mercantile concern has a large number of inverted electric-light globes under a balcony that runs around the store. Formerly it was a regular weekly job for the porter to unscrew the globes and re-

move the accumulation of dead insects and other matter that collected in the bowls. Finally, shields of glass, cut circular to fit the open ends of the bowls, and with a hole in the center to slip over the light sockets, were used to cover the openings. This arrangement did not interfere with the illumination and kept out the dust and insects. Having successfully demonstrated its usefulness, the idea was applied to all the lamps in the store.—Chas. A. Goddard, Los Angeles, Calif.

¶A small hole drilled through the socket provided in stove lids for the lifter, will allow the light from the fire to shine through so that the location of the socket is easily found in the dark.

Making Varicolored Flash Papers for Stage Effects

Amateur plays are often produced in small communities where there is no electric-lighting current. If the play calls for fire scenes, lightning, artillery fire, explosions, etc., it is difficult to produce much of an effect without electric lights. For just such occasions, a set of varicolored flash papers will produce the effects desired. These papers can be prepared at home at small cost.

It is well, at this point, to call attention to the danger attending the use of these sheets, as well as of colored fires, etc., in close proximity to scenery, costumes, and other "props" that have not been fireproofed. Sheets of tissue paper, about 12 in. square, are used; these are cut in half for small, weak flashes, while, for longer ones, a whole sheet can be used. The tissue paper is soaked in aqua fortis, which should be bought ready-mixed from the druggist. The paper is thoroughly soaked, but is taken out of the solution as soon as possible, rinsed in running water, and hung up to dry. A wooden clothespin, or something of the sort, should be used to remove the paper from the acid, which should be prevented from contact with the fingers or clothing. When the papers thus treated have dried, they are pinned to a stick, held at arm's length and touched off with a match at one corner. They will burn with a brilliant white flash.

Colored flashes are made as follows: In all cases the paper is first treated as described above, and dried before immersion in the chemical solutions that produce the colors.

For red, dip the treated sheets in a solution of water and nitrate of strontium. Blue is obtained by soaking in copper-nitrate solution. Green can be produced from a solution of copper chloride, while a solution of calcium nitrate will produce another red. A beautiful violet flash is made from a solution of saltpeter and water.

Be sure to pin the paper to the end of a stick, which can be held at arm's length, or else fasten to a tin reflector before lighting.

Dining Table Made from a Small Gate-Leg Table

Usually the extension dining table is a cumbersome affair for the small family dwelling in a modern apartment. The illustration shows a table that affords a real economy of space. A small gate-leg table was used, and provided with additional capacity for guests by using the two-piece auxiliary top shown. The thin boards used for making

Above: The Gate-Leg Table and the Two Parts of the Sectional Top. Below: The Sectional Top Hooked to the Small Table

the halves were screwed to wooden battens; hooks were provided for holding the sections together at the center and to the supporting table. The small table will serve for two or three persons, while eight can be seated when the sectional top is used.—Walter C. Harris, Brooklyn, New York.

How to Bleach Beeswax

Pure white wax, which is not only desirable for certain uses but brings a higher price on the market, is obtained from ordinary yellow beeswax. The yellow wax is sliced into thin chips and laid on cloth-covered frames or trays, supported a few inches above the ground, in the open. The flaked wax is turned over frequently, and occasionally sprinkled with clean soft water, if there is insufficient dew or rain to moisten it. The wax should be bleached in about four weeks, but if, on breaking the flakes, the wax still appears yellow on the inside, it is necessary to remelt it and expose a second time or even a third, until it is thoroughly bleached. The time required for proper bleaching depends largely upon the condition of the weather and the amount of sunshine that falls on the wax.

Shooting Gallery for Toy Pistols

Skill in shooting toy pistols, blowguns, and similar harmless weapons that use peas, marbles, or wooden darts for am-

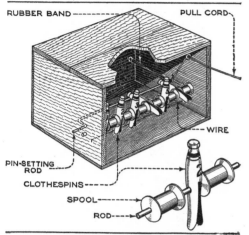

A Pistol Butt for Indoor Use, by Means of Which the Young Marksman can Improve His "Shot," with Harmless Weapons Such as Marbles, Peas, and Darts

munition, can be easily acquired by practicing on a target of the type shown in the drawing. Clothespins, spools, and some wire are about the only materials required. The clothespins, a spool between each pair, are placed on a stiff wire or small rod and fitted in a box, as shown. Back of the clothespins and a little above their lower ends is a second wire that holds them upright; this wire should be placed so that the pins will lean forward a little. When these targets are knocked over by an expert—or lucky—shot, they are caught by the pin-setting rod at the back; this rod is bent from a piece of stiff wire, and is held horizontally by a rubber band. When all the targets have been knocked over, or after each marksman's turn is over, the pins are reset by a pull on the cord tied to the pin-setting rod. If desired, the clothespins can be painted and designated by numbers.—Louis M. Steffen, Dayton, O.

How to Bake a Fish in Clay

A clay-baked fish is so simply prepared, compared with cooking it in camp, that it will be hailed with delight. The best kind of clay to use for the purpose is the gray clay generally found along streams, but in the absence of this, ordinary red clay will answer. The fish is cleaned and washed, stuffed if desired, and sewed up in the ordinary way; the head of the fish may be left on. The clay is packed around the fish so that there will be a 2-in. thickness of it at all points, and it is then ready to be tucked into the fire. Previously, a hot hardwood fire has been kept going; ash is to be preferred because it produces hot coals that last a long time. It is generally a good idea to convert the evening campfire into a heap of coals for the purpose. The fish is placed at the bottom of the coals, covered, and left overnight. In the morning the coals are scraped away, and the hard-baked clay crust broken away with the camp hatchet, exposing the thoroughly cooked fish, savory and palatable to the last morsel. A 5 or 6-lb. fish makes an ideal bake.—Robert Page Lincoln, Minneapolis, Minn.

Swing Stand Used during the Winter

A familiar type of porch swing is suspended from a frame of steel tubing so that it can be moved to any location, either on the porch or lawn. During the winter months, when the swing is not in use, the steel frame can be used to provide additional space for the storage of clothing. This is accomplished by removing the swing from its stand and hanging the clothes—which are placed on hangers, of course—from the horizontal crosspiece.—Mrs. Elizabeth Bachman, Fullerton, Pa.

A Simple Wire Stretcher

A novel arrangement, that is simply installed on any wire line to facilitate lowering or raising, or to maintain a uniform tension, is shown in the drawing. It is made from a bent bar of round iron.

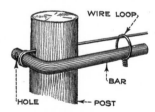

Such a device is being used, in one instance, to keep wire clotheslines taut, so that no props or other supports are necessary. The wire is led through a hole in the post and fastened through a hole drilled in the bent bar. Turning of the bar crank winds the wire around it and tightens the line. The crank is held in place and prevented from unwinding by a loose loop, or link, which is slipped over its end.

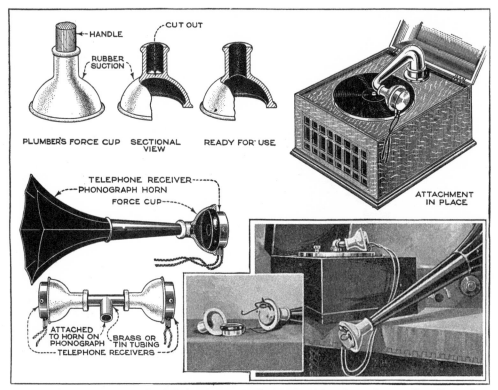

Simple Methods of Making a Loud Talker for the Radiophone-Receiving Set: One or Two Force Cups, with a Few Scraps of Tubing and a Discarded Phonograph Horn, Are the Only Materials Necessary

Making Loud Talkers for Radio-Receiving Sets

A variety of devices for amplifying the sounds received through the radiophones are on the market for the choice of the owner of a radio-receiving set—if he cares to pay the price. If his means are limited, the ideas illustrated may be applied at a cost of well under a dollar.

The only material required is one or more small force cups, such as can be bought from any plumber or hardware store. Usually these cups are mounted on a wooden handle, and this is removed, as the rubber cup is the only part used. Between the handle socket and the hemispherical cup there is a dividing wall of rubber that must be cut out, which makes the cup ready for use. One of the phone receivers is fitted inside the large end of the cup and the neck is slipped over the tone arm of a phonograph, after the soundbox or reproducer has been removed, or over the end of a horn as in the drawing. A larger volume is, of course, obtainable by using both receivers of a headset and mounting them on the ends of a tee.

Efficiency of Spark-Plug Points Improved by Sharpening

As the result of investigation as to the advantage obtained by varying the form of the electrodes through which a high-tension current was passed, it was found that the same voltage would bridge a gap from 25 to 50 per cent greater when the points were made very sharp, than when they were blunt. This knowledge was put into practice on a number of cars in which trouble was experienced due to persistent fouling of the plugs with oil and carbon. As a result of these tests, extending over a year, the invariable practice in one shop, when fitting new plugs, is to file the electrodes to a point, and make the gap the usual thickness, that of a worn dime. The practice has overcome the trouble with the plugs, and has made cylinders, which previously required weekly cleaning of the plugs, operate without a "miss" for months. It is necessary to set the points closer after a long period of operation, as the smaller points burn away quicker than the usual circular-section electrodes. It is most important to sharpen the positive electrode.

How to Make Artificial Pearls

BY R. R. HENDERSON

ARTIFICIAL-PEARL making would seem, at first glance, to be a process somewhat beyond the skill of the amateur, but, as a matter of fact, very good-looking "pearls" can be turned out in the home workshop. The art is not difficult, and can readily be acquired by practice.

For many years, artificial pearls were made by coating the outside of small glass beads with a luster paint. This method, however, had the disadvantage that constant wear caused the paint to chip off, leaving the glass bare. At the present time this method is used for the cheaper grade of goods only. The better and more expensive artificial pearls are now made by first blowing thin glass globules, with a hole in each end, from thin-walled glass tubing. Then, by a special process, the interior of the globule is coated with a luster solution.

First obtain a supply of thin-walled glass tubing of small diameter. The exact size of tubing and thickness of wall depend upon the size of pearls one wishes to make, but for most ordinary uses, tubing of $\frac{1}{8}$-in. outside diameter and walls $\frac{1}{32}$ in. thick will serve.

Place one end of a piece of tubing in the flame of a Bunsen burner, as in Fig. 1, and rotate it slowly until the glass melts and runs together, closing the hole in the end of the tube. Remove the tube from the flame and let it cool slightly. Then again insert the tube in the flame, about 1 in. from the closed end, and rotate it slowly so as to heat it uniformly all around. When the glass becomes red-hot and soft, remove it quickly from the flame and blow in the open end, as in Fig. 2, until a bulb or globule of glass appears of the desired size. Be careful not to heat the glass until too soft or to blow too hard, or the bulb may accidentally be blown so thin as to burst it. Also, if the tubing is not uniformly heated all around, the hottest spot will swell out first, and a globule of imperfect roundness will be formed. A complete globule, blown near the end of the tube, is shown in Fig. 4. After blowing the first globule, let it cool a moment so as to set and become hard. Then heat a place on the tube beside the first globule, Fig. 4, and blow another; continue in this manner until as much as possible of the length of the tube has been used, as in Fig. 5. Now, with a small three-cornered file, make a scratch between each globule and break them apart. The small tips left on each end may be ground off by rubbing on a piece of emery paper, and then polishing with wet chalk, or, preferably, they may be melted and closed in slightly to give a finer hole, by holding the globule in a pair of tweezers and rotating the end in a fine flame. Wash a quantity of the globules in a pint of water, to which a spoonful of washing soda has been added, stirring them around thoroughly; rinse them in three or four changes of clear water to remove every trace of the washing soda, then place them in a pan in an oven to dry. The globules should not be touched by the hands after washing.

Now prepare a very thin varnish by dissolving a small quantity of mastic in acetone. To this add a very little ordinary copal varnish. Then stir into this thin mixture a quantity of "fish silver." This material consists of the carefully dried, ground, and sifted white scales of fish; it may be obtained from dealers in artists' and decorators' supplies. The complete solution, already mixed up, may also be obtained, under the name of "pearl solution" or "essence d'Orient."

Construct a coating apparatus from a glass or enamel-ware jar, as shown in Fig. 6; this is so arranged that it may be rapidly rotated about an inclined axis, and should be equipped with some sort of cover. An ordinary preserve jar serves well. Place in the jar a small quantity of the fish-silver and varnish mixture, and about three times as great a bulk of the glass globules, close the cover, and rotate the jar rapidly. In a few minutes it will be found that the interior of every globule is coated with the luster mixture, and that all the excess liquid will be in the form of drops in the inside of the globules, capillary attraction, assisted by the rubbing of the globules together, which prevents the coating settling and adhering to the outside, having drawn the liquid to the inside. Now heat the jar gently with a gas flame or alcohol lamp, rotating it again rapidly, leaving the cover off. In a short time the excess liquid will have evaporated, leaving on the interior of the globules a hard coating of the luster material, and giving the effect of natural pearls. Care must be taken that the luster varnish is very thin, and that only a small quantity is used, to prevent the excess from sticking to the inside of the jar, or to the outside of the glass globules. If, by any chance, a little of the coating should stick to the outside of the globules, it may be removed by putting them

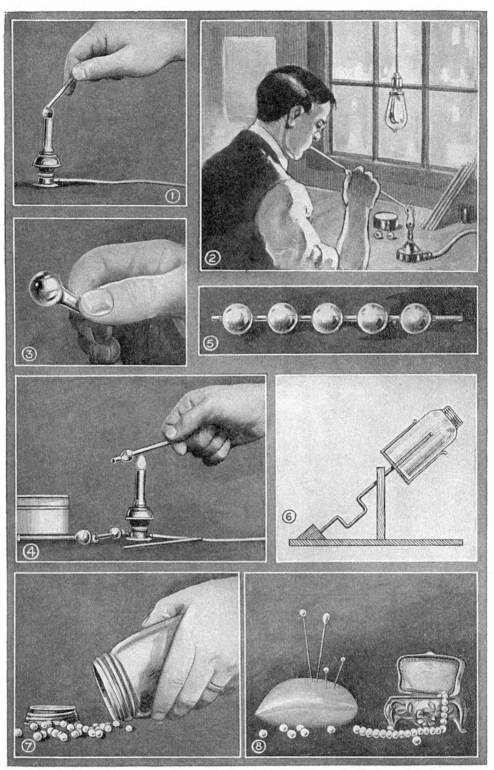

Fig. 1, Closing the End of the Tube; Fig. 2, Blowing the Globule; Fig. 3, a Hatpin Globule; Fig. 4, Preparing the Tube for the Second Bulb; Fig. 5, a Complete String of Bulbs; Fig. 6, the Coating Apparatus; Fig. 7, Emptying the Jar after Coating, and Fig. 8, Complete Hatpins, Stickpins, a Necklace, and Some Loose Pearls. This Is an Interesting Hobby, and Is Well within the Skill of the Amateur

back in the revolving jar, with an equal amount of sawdust which has been slightly moistened with acetone. If the jar is then rotated, the rubbing of the pearls against each other and the sawdust will clean them. They may be polished by rotating in the jar with fresh, clean, dry sawdust. It is best first to sift the sawdust, so that it will contain no grains small enough to get inside the pearls through the small holes. Use the sawdust that remains on the sieve.

Cork Prevents Double Exposures

After accidentally bumping the trigger of his box camera a few times and getting

double exposures, the owner decided that some step must be taken to prevent this. The method consisted simply in the insertion of an ordinary cork into the opening in front of the lens. This arrangement not only makes the opening light-tight, but permits unlimited tripping of the trigger, and is also dustproof, thus protecting the lens and shutter mechanism. It should be borne in mind, however, that the cork must not be so long as to bear against the revolving shutter disk.

An Auxiliary Seat for the Auto

Upon occasion it is desirable, especially with a two-passenger car, to accommo-

SEAT FOLDED

SEAT UNFOLDED

date an extra passenger. One method providing this extra seat is that shown in the appended sketch. This consists of a hinged seat attached to one of the right-hand doors as shown, and provided with a hook and leg on the forward edge of the seat, for its support.

To prepare glass globules for hatpins, seal one end of a glass tube by rotating it in the flame and blow the globule directly from this sealed end; Fig. 3 shows a hatpin globule. When cut from the tube by a three-cornered file, this gives a globule with only one hole. It is coated in the same manner as the ordinary beads. Pins may be fixed in place by imbedding them in a small amount of light-colored melted sealing wax run into the bead around the pin.

This seat has the advantage that it can be folded snugly against the door when not in use.

For roadside tire repairs, or other troubles, this seat provides a handy place to work without stooping, and a shelf for the tools.

Socket Wrench for Connecting Rods

The socket wrench shown in the drawing was made for the purpose of tightening the bolts used to hold the caps of automobile con-

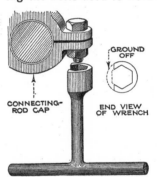

GROUND OFF

CONNECTING-ROD CAP

END VIEW OF WRENCH

necting - rod bearings. The space between the cap body and the inner edge of the nut is generally so small that only a very thin socket wrench can be used. The result is that the wrenches are often broken. For the tool illustrated, a heavy wrench is used, one side being ground off. With this wrench the nut can be drawn up tightly without resort to a hammer and chisel.

Thought Transference with Dice

A simple arithmetical trick, that is very impressive, consists in asking a member of the party to throw a pair of dice on a table, then telling him what he has "thrown." In order that the person performing the trick may not see the dice, he goes to another part of the room, where he remains until the conclusion of the trick. The performer then requests the person casting the dice to use the uppermost figure on one of the dice and multiply this by 2, then add 5 to the product, multiply the sum by 5, and, finally, add the uppermost figure on the second dice. When he has done this, the performer tells the number of spots showing on each of the dice, although he has not seen them.

When the performer is told the number that has been obtained by the above process, it is only necessary for him to subtract 25 from it, and the remainder will be a two-figure number; one of these figures will be the number of spots showing on one of the dice, and the second will be that on the other. Thus, for example, assuming that the person tells the performer that his final number is 67, by subtracting 25 from it, 42 is obtained; then the performer knows that the figure on one of the dice is 4 and that on the other is 2. The operation from beginning to end is: $4 \times 2 = 8 + 5 = 13 \times 5 = 65 + 2 = 67 - 25 = 42.$—Raymond Dixie, New Haven, Conn.

Reading Lamp Made from Music Rack

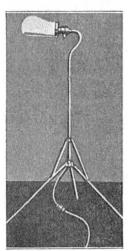

An old music stand can be made into a very serviceable and satisfactory adjustable reading lamp simply by attaching a lamp to it. The flattened end of the sliding tube is sawed off, and the tube filled with melted resin and bent as shown, after which the work is heated in order to melt the resin and allow it to run out of the tube. This makes a reading lamp that can be adjusted for height and is readily portable.—L. Ringer, Cincinnati, Ohio.

A Camp-Site Incinerator

Proper sanitation is a first necessity in an auto-tourist's camp site. A small garbage incinerator can be built at little expense, and the aid it will render in keeping the place sanitary is well worth the small outlay of labor and materials. The inner lining should be of fire brick, while ordinary bricks, old or new, will serve for facing. Built like an old-fashioned beehive, the incinerator has a fire grate about 18 in. from the bottom,

Incinerator for a Camp Site Used Mainly by Auto Tourists Helps to Keep the Camp Sanitary and Free of Refuse of All Kinds

with an opening at the top equal to about one-half its diameter at the bottom. After a wood fire has been started under the grate, the refuse is dumped through the top hole, and will be gradually consumed without further attention on the part of the caretaker. Further to facilitate the disposal of refuse, each tourist, as he enters the camp, should be provided with a wire basket in which he is requested to deposit the refuse from his tent.—C. L. Meller, Fargo, N. D.

Sewing Kit Made from a Cigar Box

An ordinary cigar box, preferably one used for holding 50 cigars, provides a good container for spools of thread, and other sewing materials, when fitted up as shown in the illustration. Two long aluminum knitting needles, such as can be obtained for a few cents, are fitted into holes drilled in the ends of the box. In the end at which the

knitting needles are inserted, the holes are a loose, sliding fit, while at the opposite end they are a little smaller so that the needles will be held in place by friction. The spools turn freely on the rods, which need never be removed except when an empty spool is taken out to make room for a new one. The remaining space in the box can be used for pincushions, scissors, and other articles.

A Novel Photo-Print Washer

A great deal of time is saved by the user of the print washer illustrated, and the prints are thoroughly washed, as they do not come into contact with each other, and each is surrounded by circulating water. Ordinarily, the prints are removed from the hypo fixing solution, dumped into a pan or other vessel, and the water turned on; as the wet prints always stick together more

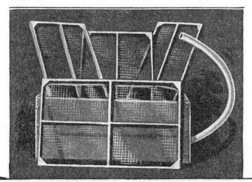

copper-screen wire, to prevent rusting, and the sides, ends, and partitions are made of narrow strips of galvanized iron, soldered together and to the wire screen. A small strip of wood is placed in the bottom of the box to keep the trays from the bottom, and allow a free circulation of water underneath. The hose that carries the water from the faucet to the washer enters the box at the bottom. In use, a strip of sheet metal is placed over the top and fas-

 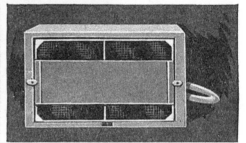

Above: The Wooden Container and Four Copper-Wire Trays, Each Divided into Four Compartments. Left: Top View of the Washer. Right: Strip of Sheet Metal Fastened at Each End over the Top of the Washer

or less, there is no assurance that all of the prints are thoroughly washed.

This print washer consists of a wooden box, which serves as a container for four trays, divided into four compartments, one for each print. The trays are made of

tened with buttons on each end of the box, to prevent the prints in the top tray from washing out. Double the number of prints can be washed by placing two, back to back, in each compartment.—John T. Daniel, San Francisco, Calif.

The Care of Leather-Bound Books

Bookbinding leathers have a tendency to dry out and finally break at the hinge, if the leather is not kept flexible by occasional greasing. For this purpose any high-grade vaseline, free from acids, may be used. The following method of greasing the leather is that used in a large library to preserve the sheep bindings of legal volumes:

The work should be done in a well-lighted, airy place, preferably in the summer, when the windows can be opened to let in the outside air. Place the book, back up, on a table having a smooth top, and coat the back with vaseline, rubbing it well into the grain of the leather with the hand. Next, treat the sides and edges, being careful not to get

any of the vaseline on the paper. Use small quantities of vaseline and rub it in well, with a firm, flexible stroke. The first application is usually absorbed rapidly. Next apply a second coat and rub the leather down well as before. The back and back edges require more vaseline than the sides and front edges. Then place the book on a shelf to dry, which takes from 24 to 48 hours, and finally rub off all surplus grease.

Merely dabbing on the vaseline with a cloth and rubbing it off does not suffice. Neither is one application, left to dry in by itself, satisfactory. Light rubbing with cloth pads will not do the work as well as rubbing with the bare hands. Also it is better to use small quantities of vaseline, and make several applications, according to the condition of the book.

Fusible Alloys for Setting Crystals

Radio-sensitive crystals, such as galena, silicon, and the like, should be imbedded in metal to obtain the best results, and this is done by using an alloy that melts at a point considerably below the boiling temperature of water. Heat destroys the sensitiveness of the best crystals, and for that reason lead cannot be used.

An alloy which melts at 197° F. is composed of lead, 3 parts; tin, 2 parts, and bismuth, 5 parts. The melting point of the metal can be still further reduced by adding 1 part of warm mercury to the molten alloy when it is removed from the fire. The addition of mercury will cause the alloy to remain liquid at 170° and become a firm solid only at 140°. The boiling point of water is generally taken as 212° F., although the higher the elevation above sea level the lower this temperature will be.

Making a Toy Catapult

A 10-cent rat trap of the type shown in the drawing can easily be made into a marble-throwing catapult, the range of the missile being regulated by an adjustable stop. The trap is fastened to the edges of the ammunition box and the bait hook is removed. The stop is then bent from a strip of sheet metal and fastened to opposite edges of the trap as indicated. Two side arms that serve as braces for the stop are adjusted by means of a wire pin passing through holes in the stop and arms. The throwing arm should be made of ½ by ½-in. hardwood, about 10 in. long, although the length is best determined by trial. A small metal cup at the end of the arm provides a pocket for the ammunition. If desired, a trigger arrangement, similar to that shown, can be added to the device. Flour, tied in tissue paper, may be used to make a realistic bomb, as it gives off a smokelike puff when it strikes, and is harmless. The longest throw the device is capable of, will usually be attained when the stop is set at an angle of about 45 degrees.—J. H. Kindelberger, Cleveland, Ohio.

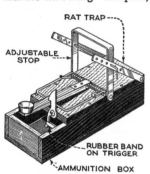

Straightening a Bent Auto Axle

When an automobile skidded into a partly obscured curb, one of the rear-axle shafts was badly bent. The wobble of the wheel was so pronounced that further driving was out of the question. This happened near the garage, and as the use of the car was necessary for delivery purposes, the method illustrated was used to straighten up the wheel. A piece of plank, about 3 ft. long, had a hole cut in its center to clear the wheel hub. Driving the car into the doorway of the brick garage, a jack was placed between the upper end of the plank and the side of the doorway. The opposite wheel was blocked to prevent the car from being pushed sidewise, and pressure exerted on the jack straightened the axle. Repairs made in this manner should not be considered permanent.

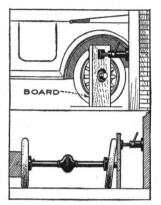

Windmill Controlled from the House

Having to go from the house to the windmill, some 30 rods away, in order to stop it when the tank is full, soon becomes quite a nuisance. To avoid this, and to control the operation of the pump from the house, the arrangement shown by the drawing was worked out. A bell crank made from a piece of tire iron, was pivoted to one of the tower legs with a bolt. Wires fastened to the ends of the crank were led to the house and to the controlling mechanism of the windmill for throwing the pump out of gear. The pump can also be started by releasing the control wire.—Rufus E. Deering, Ottawa, Kan.

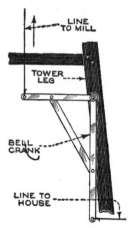

Wheatstone Bridge for Coil Winding

To wind electrical coils that must possess a certain resistance, or to wind a coil that will have the same resistance as another, are real problems to the experimenter unless he knows how to go about it. Almost every amateur is familiar with

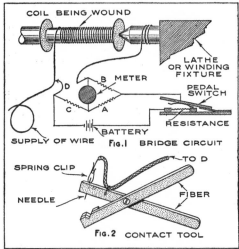

SUPPLY OF WIRE   Fɪɢ.I   BRIDGE CIRCUIT

Fɪɢ.2   CONTACT TOOL

Winding a Coil of Wire to a Certain Resistance Is an Easy Task When the Wheatstone Bridge Illustrated is Used

the Wheatstone bridge, by name at least; this is usually represented by four resistances in a diamond-shaped figure, with a battery connected at two of the points of the diamond and a galvanometer across the other two. When there is no current passing through the meter, it shows that the four resistances are balanced, and in proportion; that is, A is to B as C is to D. The voltage of the battery and the markings on the instrument need not be known or correct, as the accuracy of the result does not depend on them. Fig. 1 shows a method of applying the bridge principle to a coil as it is being wound, so that one may know just where to cut off the wire to get the right resistance. The galvanometer should, of course, be as sensitive as possible; it is not difficult to make one from a compass and a coil or two, or an excellent galvanometer can be obtained by removing the resistance coil from any good moving-coil voltmeter, or by connecting direct to the meter so as to bypass the resistance. The resistances A, B, and C must be of known value, if the coil is to be made a definite number of ohms, but if any one of them is of known resistance, two more can be made equal to it, by using ordinary bridge

methods, so that only one known resistance is necessary at the start.

In the figure, the coil being wound forms side D of the bridge; the inner end, or that next the meter, is grounded on the dead center of the lathe or winding fixture, and a temporary contact is made with the last turn wound by piercing the insulation. The pedal switch shown serves two purposes: It connects the battery in the circuit after the contact is made at D, and, as the pedal is further depressed, shorts out a resistance. This resistance is in series with the battery, and should be high enough to protect the meter from the high current which will result if the four sides of the bridge are much out of balance. When the first contact of the pedal switch closes and it is seen that only a small deflection appears on the meter, then the pedal can be pressed farther down, shortening out the resistance and giving greater accuracy in balancing the coil resistance against resistances A, B, and C. Fig. 2 shows a simple and handy tool for making the contact at D. Two pieces of fiber or hardwood are pivoted together at the middle; one is notched, and through the other a sharp steel needle is driven which is connected to the wire leading to the bridge. By laying the magnet wire in the notch and gently closing the needle point against it, the point will pierce the insulation sufficiently to make contact, but without damage of any importance to the cotton or enamel insulation.

Spreader for Split Rims

Automobile owners and drivers generally have trouble in bringing the ends of a split rim together when a new tire has been placed on it. Various tools are made for this work, but unfortunately, they are not usually carried on the car. The drawing shows a simple method of using the jack, which no motorist is without, and a short block of wood.

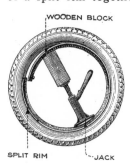

Pressure exerted on the rim through the jack and block will force the ends together speedily and easily. A block may be placed under the jack, to protect the rim.—Harry G. Schultz, Teaneck, N. J.

Two Effective Card Tricks

The first trick involves the use of four cards, which are "fanned" out, to show a corresponding number of kings, the performer repeats the magical "abracadabra," and, presto! the same hand has changed to four aces when it is again displayed—a third pass, and only blank cards are shown. Six cards are required for this trick, three of which are unprepared, the other three being "prepared." The three unprepared cards are the king, ace, and blank card shown in Fig. 1, the three other cards being prepared by pasting a part of the remaining three kings over a corner of the aces of their corresponding suits as shown in Fig. 2. In the presentation of this trick, the four kings are first displayed to the audience. The real king being on top, the cards are fanned as in Fig. 3, so as to show only the kings on the corners of the other three cards. Then, the performer picks up the ace of spades, which has been left face up on the table, and announces that he will place it directly behind the king of spades, which he does. He then lays the king of spades on the table. The cards are then closed up and turned over so that the cards are held at what is the top of the cards in the first presentation of the four kings. Then, the cards are fanned out to show the four aces, as in Fig. 4. The index numbers in the corners of the aces should be erased or covered up, otherwise it will be impossible to show the blank cards.

The manipulator now states that by placing a blank card, which he picks up from the table, where the ace of spades is, the spots will disappear from all the cards. The ace of spades is placed on the table, the blank card taking its place. The cards are then closed and fanned out, the hand showing four blank cards, as in Fig. 5. In the second trick, an ace of diamonds is held in one hand and an ace of spades in the other, but while held in full view of the audience, the cards change places. The prepared cards are made from two aces of diamonds, from which the corner index pips and letters have been erased. An ace of spades is also required, the center of which is cut from the rest of the card as indicated in Fig. 6, which shows the appearance of the

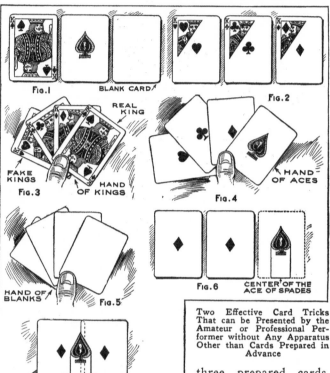

Fig.1 BLANK CARD Fig.2 REAL KING FAKE KINGS Fig.3 HAND OF KINGS HAND OF ACES Fig.4 HAND OF BLANKS Fig.5 Fig.6 CENTER OF THE ACE OF SPADES THE FAKE SPADE IS SLIPPED FROM ONE CARD TO ANOTHER Fig.7

Two Effective Card Tricks That can be Presented by the Amateur or Professional Performer without Any Apparatus Other than Cards Prepared in Advance

three prepared cards. In presenting this particular trick, an ace of diamonds is held in each hand, but only one of them is visible to the audience, the other being concealed underneath the ace that has been cut from the card. The performer then announces his intention of making the cards change places. He turns the backs of the cards toward the audience, and, with the hands apart, begins moving the cards back and forth, bringing them a little closer to each other at each pass. Finally, when the edges touch, as in Fig. 7, the false center from the card is slipped over and onto the other card; this done, the cards are moved back and forth, gradually separating them, and their faces are again turned to the audience, when, to all appearance, the cards have changed their positions.

How to Make an Inlaid Checkerboard

By EDWIN M. LOVE

IN the checkerboard design illustrated, each square is a miniature checkerboard in itself, composed of 64 light and dark squares.

For the dark squares, 16 strips each of ebony and mahogany, ³⁄₁₆ by ½ by 17 in., are needed, and as many strips of maple and oak of the same size will be needed for the white squares, a total of 64 strips. Using alternate strips of light and dark-colored wood, eight strips are glued together to make a laminated piece, 1½ in. wide. From this piece, after it has been glued and both faces sanded, ³⁄₁₆-in. sections are

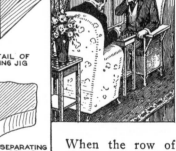

METHOD OF PREPARING LAMINATED STOCK

WHITE OAK — MAPLE — MOVABLE BLOCK — GUIDE — STOP

MAPLE / WHITE OAK / MAHOGANY / EBONY

DETAIL OF GLUING JIG

SEPARATING STRIPS — BACKING BOARD

¼" PROJECTION

METHOD OF ASSEMBLING BLOCKS

An Inlaid Checkerboard, the Squares of Which Are Miniature Checkerboards Themselves, being Built Up of Light and Dark-Colored Woods

cut transversely, and eight such sections glued together to form a square of the board, using the gluing jig illustrated for the purpose. The sections are cut ³⁄₁₆ in. wide in a miter box. The strips are glued together so that light squares will come next to dark ones, the mahogany and ebony being used for the "black" squares, and the oak and maple for the "white" ones.

All of the gluing operations should be done in a warm room and the stock should also be warmed. The wooden gluing jig should be made long enough to accommodate about eight squares at a time, each being separated from the other by a strip of paper. By rubbing the edges of the jig with paraffin, the glue will be prevented from sticking, and the blocks can easily be removed.

For a backing board use a ½-in. pine board, 14½ in. square. Nail and glue two ¹⁄₁₆ by ½-in. red-oak strips near two edges, to form a right angle, and also prepare

seven strips of red oak, ¹⁄₁₆ by ½ by 12½ in., and 56 similar strips, 1½ in. long. Paint the face of the board with glue and place in the angle formed by the strips at the edge, which should be at the lower left corner, one of the dark squares, with an ebony square in the corner, and the grain running parallel with the left edge of the board.

Glue one of the short separator strips to the upper edge, and then place a light block in position, with the grain at right angles to that of the dark, and with the light maple square in the lower left corner.

When the row of eight squares has been placed against the pine strip at the left, glue one of the long separator strips to the inner edge of the row and start building up the next row. When all the blocks are assembled, inclose the two open sides with strips, and clamp them tightly until the glue sets.

When the glue has dried thoroughly, trim the ends of the separator strips flush, and glue ¹⁄₁₆-in. strips around the remaining two sides. Cut off the edges of the backing board to leave a ¼-in. projection, and scrape and sand the surface smooth; wax it, or, if a higher gloss is desired, fill with a light-paste filler and give three coats of varnish.

❡ Before using raveled-out wool for knitting, wind it in hanks, and place it in a steamer until thoroughly moistened. Then let it dry and wind into balls. The yarn will be straightened out and the color freshened.

Safety Spring-Hanger Bolts

One job that will try the patience of the automobile repairman consists of lining up the eye of a spring with the spring hanger in the chassis. The stiff springs are difficult to handle unless some means is provided for alining the two eyes. A simple jack for flexing the spring and lifting the weight of the car can be improvised from a bolt and a short piece of pipe or tubing, as indicated in the drawing. These materials are usually easily obtainable, and with a few of these jacks, much time can be saved in working the eye of the spring into the proper position for the insertion of the shackle bolt.—G. A. Luers, Washington, D. C.

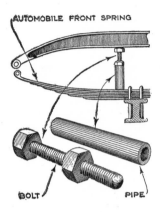

AUTOMOBILE FRONT SPRING

BOLT PIPE

A Simple Dimming Switch

The drawing shows a very sensitive and simple dimming switch for use with an electric incandescent lamp, in locations where a dim light must be kept burning constantly. A hard-fiber, or bakelite, rod is turned down, as shown, to form a plunger, leaving enough of the original diameter to form a head, which is then knurled. A portion of the length of this rod, at the upper end, is threaded to fit the tapped hole in the mercury container. The latter consists of another piece of rod which is bored and tapped, the diameter of the hole below the thread being slightly larger than the plunger. The plunger is drilled, along the vertical axis, with a $\frac{1}{16}$-in. drill, a copper wire being run through the hole, and projecting through the bottom of the plunger. Resistance wire is then wound around

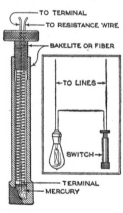

TO TERMINAL
TO RESISTANCE WIRE
BAKELITE OR FIBER
TO LINES
SWITCH
TERMINAL
MERCURY

the outside of the plunger, being firmly fastened at the bottom and led in at the top through another $\frac{1}{16}$-in. hole paralleling the center one. At the bottom of the container, a small quantity of mercury is placed; as the container is screwed upward, the mercury rises in the annular space between plunger and container, so that the current flows through less of the resistance wire, thus increasing the current strength until full light is obtained. Screwing the container downward, of course, reverses this process. This type of switch has been successfully used on small motors and baking ovens, in addition to lamps.—W. Burr Bennett, Honesdale, Pennsylvania.

Combined Spring Spreader and Tire Remover

A tool that serves equally well for spreading the leaves of automobile springs for the application of lubricant, or for removing a tire when the beads have rusted to the rim, is shown in the drawing. The tool can be forged on the anvil, from parts of old wagon tires, in less than an hour.

A shouldered rivet is permanently fastened to the curved part, as illustrated, and curved slots that fit around the shoulder of the rivet are provided in each end of the lever handle, but on opposite sides. The span of the tool is somewhat different for the two uses, and for that reason the blade ends are not quite duplicates, the difference being in the distance from the pivot hole to the end of the blade. The hook portion of the tool is simple enough to understand without explanation, while the only detail of the handle that need be further explained is that the end used for tires should be blunt, the opposite end being sharpened for easy insertion between the spring leaves. The blade can be removed from the device and used as an ordinary tire tool for pulling the bead over the rim.

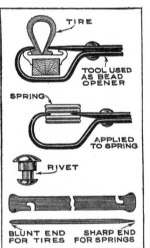

TIRE
TOOL USED AS BEAD OPENER
SPRING
APPLIED TO SPRING
RIVET
BLUNT END FOR TIRES SHARP END FOR SPRINGS

Spring Prevents Breaking of Fishline

"The big one that got away" would, perhaps, have been landed, had the strain on the line been applied more easily when the fish struck the bait. However, as fish are not particularly considerate, the angler must provide a means for preventing his line from breaking, and one that will materially assist in landing his game. This may be done by merely interposing a short length of small-diameter spiral spring of the proper tension between line and hook. The spring will absorb a considerable portion of the strain on the line caused by the vicious lunging of certain varieties of game fish.—O. S. Billings, Ruskin, Fla.

A Jumping-Frog Toy

An entertaining little toy can be made from the wishbone of a fowl after it has been well cleaned and freed from flesh.

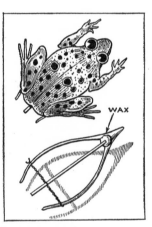

Take a piece of strong, thin string and double it, tying it securely to opposite sides of the wishbone about 1 in. from the ends, as in the drawing. Cut a strip of wood a little shorter than the bone, and make a circular notch about ½ in. from one end. Push the stick through the doubled string for about half its length, twist the string tightly by means of the stick, then pull the stick through until the notch is reached. From a piece of paper or thin cardboard cut out the outline of a frog, paint it to resemble the animal as nearly as possible, and paste this to one side of the wishbone. The only thing now required is a piece of shoemaker's wax, which is placed on the underside of the bone, just where the free end of the stick will rest. When it is desired to make the frog jump, push the stick down and press the end into the wax. Place the frog on the table, and after a short while the toy will, all of a sudden, make a very lifelike leap as the end of the stick pulls away from the wax.—S. Leonard Bastin, Bournemouth, Eng.

The "X-Ray" Pack of Cards

This trick is a "mind-reading" stunt which is worked on a new principle, and is very puzzling. A full pack of cards is shown and half of them are handed out, the other half being kept by the performer. A spectator is asked to select any card from those he holds, and insert it in the pack held by the performer, while the latter's eyes are closed or his head is turned. Without manipulating the pack in any way, the performer places it against his forehead and instantly names the card chosen by the spectator.

The cards held by the performer are prepared for the trick by cutting a slot ⅛ in. wide and 1 in. long in one corner of 25 cards, with the sharp point of a penknife, in such a manner that all the slots coincide. In presenting the trick the performer keeps all the prepared cards, and also one card which has no slot, the latter being kept on top of the pack so that the slots cannot be seen by the spectators. The performer's thumb is held over the slots when the bottom of the pack is shown. A spectator is asked to insert a card face down into the cards the performer holds in his hand. When this is done the thumb is lifted from the slot as the cards are raised to the forehead, when the performer can look through the pack and see the index on the card the spectator has selected. After the forehead "stall," the performer announces the card selected. The trick is repeated by "fanning" out the cards and extracting the card named.

Corked Bottle Used as Float

A simple and effective substitute for a wooden float, which becomes waterlogged after continued use, for water tanks and similar purposes, can be made from old bottles. The bottle is stoppered with a tight-fitting cork, which is made leak-proof with sealing wax. The bottle can be fastened to the end of a float arm by means of metal clamps.

Artistic Lamps Made from Vases and Drain Tile

There are few household furnishings that create an atmosphere of hominess

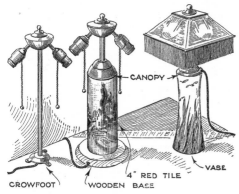

CANOPY — 4" RED TILE — VASE — CROWFOOT — WOODEN BASE

Attractive Table Lamps That can be Made from Such a Humble Article as the Ordinary Brick-Red Farm Drain Tile, or from Vases of Any Size and Form

and furnish cheerfulness to the whole family more completely than lamps which harmonize with their surroundings.

The common brick-red drain tile can be elevated to a position of beauty by using it as a part of a lamp of the type shown in the drawing. The tile itself can be decorated in any desired manner, as by painting some simple scene in oils against an enameled ground. The wooden base should have a diameter large enough to prevent the completed lamp from being easily upset. The electrical fittings, shown at the left, can be obtained from any dealer in such articles. The crowfoot should be screwed to the center of the wooden base in such a manner that the upright pipe will be in the exact center of the tile. When all parts have been assembled the space inside the tile is filled with cement. The vase lamp is practically the same in construction, but because the value of the vase is to be retained, no hole is drilled in its base, which makes it necessary to run the cord in over the edge, and down to the crowfoot. In order to make a lamp of this type stable, the vase is partly filled with dry sand.—Truman R. Hart, Ashtabula, Ohio.

Soil Crusher Made from Mower Wheels

A very useful implement for the farm, for breaking up hard lumps or clods on the surface, is a soil crusher made from old mowing-machine wheels, or wheels from a similar implement. If the maker of such a device does not possess a sufficient number of wheels, a few visits among his neighbors, or to the nearest junk man, will provide plentiful material.

The construction of such a roller is quite simple and is done by merely mounting the wheels on a shaft of the proper length and diameter, with a thin washer between each pair of wheels. This shaft is mounted on a strong wooden framework of 4 by 4-in. lumber. A tongue is attached to the implement for hitching to the horses or tractor. The wheels revolve freely on their stationary axle and consequently the proper differential effect is obtained so that corners can be turned without skidding. The wheels should be kept well lubricated to prevent undue wear on the shaft.

Hook Pulls Straw from Stack

Looking somewhat like the harpoon of the old-time whalers, the wooden hook shown in the drawing is used for pulling hay or straw from a stack, without wasting a considerable amount of it. The hook can be made from either wood or metal, so that it can be thrust easily into the stack. The barb is formed about 3 in. from the end; this

HICKORY HOOK

catches the straw, and, when the hook is pulled out, a quantity of straw comes along with it. By using this method it is unnecessary to disturb the top that has been arranged to shed the rain. In case a metal hook is used, it can be fastened to a wooden handle, but a straight sapling will answer just as well, especially if there is a conveniently formed crotch near one end that can be utilized in forming the barb.

Flexible Liquid Court-Plaster

A formula for making a flexible liquid court-plaster that can be painted over small bruises, cuts, and other unimportant wounds to keep them free of dirt, is made in the proportion of 10 oz. of flexible collodion, which can be obtained at any drug store, 20 gr. of Canada balsam, and 10 drops of castor oil.